JOHN A. BEEKMAN

Two Stochastic Processes

Two Stochastic Processes

By

JOHN A. BEEKMAN

Ball State University

ALMQVIST & WIKSELL INTERNATIONAL
Stockholm – Sweden

A Halsted Press Book

JOHN WILEY & SONS
New York · London · Sydney · Toronto

© 1974 Almqvist & Wiksell International AB, Stockholm

All rights reserved

Published in the U.S.A., Canada and Latin America by Halsted Press,
a Division of John Wiley & Sons Inc., New York

Library of Congress Catalog Card Number 73-21380
Halsted Press ISBN 0 470-06175-8

Printed in Sweden by
Almqvist & Wiksell, Uppsala 1974

*To the Memory of
My Brother Bill*

PREFACE

This book is devoted to the theory and applications of Gaussian Markov and collective risk stochastic processes. These two classes of stochastic processes began the subject of stochastic processes. L. Bachelier and Albert Einstein began the treatment of one Gaussian Markov process in 1900, and 1905 respectively and F. Lundberg did likewise for one collective risk process in 1903. (References are given in Chapter 1.) These two classes of stochastic processes have remained very important in the development of the theory and applications of random processes. This book involves applications in insurance (collective risk), physics (quantum mechanics), electrical engineering, and statistics (limit laws for Kolmogorov and Kac statistics). The book omits large segments of the topics and I apologize to those authors whose equally important contributions are omitted. However, the difference of topics emphasized is the most important reason for writing another book on stochastic processes. Although some of the topics covered are drawn from my research, the main ideas are those of other and more well-known researchers.

This book was written (and frequently revised) with two audiences in mind. Actuarial students of risk theory, for example those studying for Part 5 of the exams of the Society of Actuaries, can profitably study Chapter 1, Chapter 2, Chapter 3, and possibly sections 4.0 and 4.7 of Chapter 4, and Chapter 6. The book may be used in a probability and stochastic processes course as a supplement to another textbook. For example, I have so used the book the past three years, with the major textbook being William Feller's *An Introduction to Probability Theory and Its Applications*, Volume 1, 3rd Edition, Wiley, New York, 1968. For these students, Chapters 1, 2, 3, the above mentioned parts of Chapter 4, and Chapter 6 are fairly well self-contained. The rest of Chapter 4, and Chapters 5 and 7 are more useful as reference material, especially for those with a measure-theoretic background. It is also hoped that this book will serve as a useful reference book for researchers in collective risk, quantum mechanics, electrical engineering, and statistics.

There are numerous problems scattered throughout the book. Complete or partial solutions are given in the back of the book for many of the exercises. It is hoped that the questions and answers on collective risk will be especially helpful, as the supply of published problems in this area is limited. The first two parts of the first chapter are motivational, very heuristic, and should be read for amusement only by anybody with a little knowledge of stochastic processes.

Comments of readers will always be appreciated.

Acknowledgements

Much of this book was written at The University of Iowa during 1969–1970 while on sabbatical leave from Ball State University. Parts of the book were used at Ball State University during 1970–1973, and revisions were made in Iowa City during the summer of 1971. The comments of my students led to many improvements in the original version. Dr. Alan Huckleberry used parts of the book in one of his classes at Notre Dame University, and made several valuable suggestions. Dr. Hans Gerber, and Dr. Wan Joon Park also contributed helpful suggestions. Considerable encouragement was given by Professor Harald Cramér, Professor Takeyuki Hida, and Mr. David Halmstad. I wish to thank Professor Robert Hogg and his colleagues at The University of Iowa for their warm hospitality and helpful discussions. I also wish to thank the National Science Foundation for their eight years' support of the basic research which led to writing this book. Finally, I am indebted to my wife, Jane, for her excellent typing of the first draft of this book, and to Debbie Webb for her painstaking typing of the final version of the book.

Muncie, Indiana

John A. Beekman

CONTENTS

CHAPTER 1. Informal Remarks about Stochastic Processes 11
  1.1. The Probability of Controlled Fission 12
  1.2. A Tobacco Industry Example 16
  1.3. Noise Calculations 17
      References 20

CHAPTER 2. Some Mathematical Preliminaries 21
  2.0. Summary 21
  2.1. Historical Remarks on Integration Theory 21
  2.2. Definition of a Stieltjes Integral 21
  2.3. Existence of the Stieltjes Integral 22
  2.4. Properties of Stieltjes Integrals 22
  2.5. Easy Methods of Calculating Stieltjes Integrals 23
  2.6. Applications of Stieltjes Integrals to Probability 26
  2.7. Laplace and Laplace-Stieltjes Transforms 31
      References 34

CHAPTER 3. Collective Risk Stochastic Processes 35
  3.0. An Overview of the Subject 35
  3.1. Approximations to $F(x, T)$ 57
  3.2. A Convolution Formula for $\psi(u)$ 67
  3.3. A Monte Carlo Approach to $\psi(u, T)$ 71
  3.4. A Moment Approach to $\psi(u, T)$ 73
  3.5. Inverting Transforms to Obtain $\psi(u, T)$ and $\psi(u)$ 74
  3.6. Net Stop-Loss Premiums 76
  3.7. A Measure for the Collective Risk Process 79
      References 84

CHAPTER 4. Gaussian Markov Stochastic Processes 87
  4.0. Introduction 87
  4.1. Function Space Integrals 94
  4.2. Sequential Integrals 103
  4.3. Gaussian Markov Processes, Partial Differential Equations,
      and Integral Equations 108

4.4. Relation between Wiener and Gaussian Markov Integrals . . . 118
4.5. Monte-Carlo Approximation of Conditional Wiener Integrals . 119
4.6. Radon-Nikodym Derivatives of Gaussian Processes 121
4.7. Useful Distributions 125
     References 129

CHAPTER 5. Connection between the Two Processes 131
     References 138

CHAPTER 6. Applications to Statistics 139
6.1. Kolmogorov Statistics 139
6.2. Kac Statistics 142
     References 154

CHAPTER 7. Applications to Physics—Feynman Integrals 155
7.0. Introduction 155
7.1. Sequential Feynman Integrals 157
7.2. Generalized Schroedinger Equations 162
7.3. Approximations to Feynman Integrals 165
7.4. Analytic Feynman Integrals and Examples 168
7.5. Forward Time Equations 172
7.6. Capsule View of Some References 174
7.7. Finite Difference Approach to Generalized Schroedinger Equations 176
     References 179

ANSWERS TO EXERCISES 181

CHAPTER 1

INFORMAL REMARKS ABOUT STOCHASTIC PROCESSES

The subject of stochastic processes arose from the desire to build mathematical models for certain natural processes. The random movements of small particles (called Brownian motion) were analyzed mathematically by L. Bachelier [1] as early as 1900 and applied to the French stock market. Albert Einstein called the attention of physicists and other researchers to the mathematics of Brownian motion in his 1905 work [3]. The study of Brownian motion was greatly enhanced by the work of Norbert Wiener beginning in 1923 [10], and hence the stochastic model is usually called the Wiener stochastic process. This process has played an important role in quantum physics, and in some statistical problems. Actuaries can take pride in the early recognition (1903) by F. Lundberg [8] of the value of looking at the ensemble of risks. His papers began the study of the collective risk stochastic process.

There has been much written about the creation and use of mathematical models. A notable quote by Mark Kac appears on page 699 of [7]: "Models are, for the most part, caricatures of reality, but if they are good, then, like good caricatures, they portray, though perhaps in distorted manner, some of the features of the real world. The main role of models is not so much to explain and to predict—though ultimately these are the main functions of science—as to polarize thinking and to pose sharp questions."

The subjects of probability and statistics consider one, two and possibly many random variables. A random variable is a function whose domain of definition is a sample space and whose range is usually a subset of the real numbers.([1]) We frequently denote the sample space by Ω and its points by ω's, and the random variable at a point ω by $X(\omega)$. It is of considerable interest to assign weights to sets of the following form: $\{\omega: X(\omega) \leq \alpha\}$ for each real α. We will assume that with the given sample space Ω, there was given a probability function which assigned numbers to subsets of Ω. We will refer to the family of subsets of Ω as B. We use that function,

---

[1] A formel definition of a random variable is given in section 2.6.

say P, to assign the desired weight to the above set. We define the distribution function of X to be:

$$F(\alpha) = P\{\omega: X(\omega) \leqslant \alpha\}, \quad -\infty < \alpha < \infty.$$

Because we assume $P(\varphi)=0$, and $P(\Omega)=1$, we obtain $F(-\infty)=0$, $F(+\infty)=1$. The distribution function is also non-decreasing.

It frequently happens that more than one random variable is defined on the same sample space (Ω, B, P), each with the same distribution function. It then is appropriate to say that they are *identically distributed*. Furthermore, many real-life situations dictate that the random variables are independent which means, for example, that the measure we assign to the set $[\omega: X_1(\omega) \in A, X_2(\omega) \in B]$ is $P[\omega: X_1(\omega) \in A]P[\omega: X_2(\omega) \in B]$. Functions of one or more random variables are new random variables and their distributions are of interest. For example, the distribution of the sample mean $\bar{x}=(X_1+X_2+...+X_n)/n$ is discussed in Hoel, [6], on pages 121–127. The distribution of the maximum of X_1 and X_2 could be evaluated, as another example. The study of stochastic processes involves infinite collections of random variables. Sometimes these are indexed by the integers, that is, $X_1, X_2, X_3, ...$. More frequently, they are indexed by a parameter in an interval of numbers. Thus $\{X_t, 0 \leqslant t \leqslant T\}$ can be used to denote an infinite collection of random variables, one for each point between 0 and T, inclusive. The interval $0 \leqslant t \leqslant T$ may represent time. Prime attention is given to calculating distributions of new random variables defined as functions of all the random variables. For example, if $\text{maximum}_{0 \leqslant t \leqslant T} X(t)$ denotes the largest of the $X(t)$'s between 0 and T, an interesting problem is the determination of the distribution $P[\text{maximum}_{0 \leqslant t \leqslant T} X(t) \leqslant \alpha]$.

We will now describe some examples from physics, statistics, and engineering which will help motivate the later chapters. The examples have been greatly simplified and we do not suggest that they are accurate descriptions of real problems.

1.1. THE PROBABILITY OF CONTROLLED FISSION

Practically everyone has observed dust particles rise and fall by sitting in an easy chair close to a small slit of sunlight with the door ajar.

The same is true of small particles suspended in fluids or gases. This is usually referred to as Brownian motion because this erratic movement was first observed in 1827 by an English botanist, Robert Brown.

Consider paths of these particles

The position of a particle at any time t_0 would be an ordered triple $[x(t_0), y(t_0), z(t_0)]$ where $x(t_0)$ would be the first coordinate, etc.

For ease of thinking, let us now focus on the first coordinate, and assume that at $t=0$, $x(0)=x_0$ a constant.

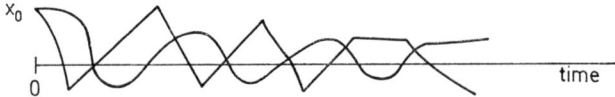

For any t, the position of the particle is random.

Let $X(t)$ represent the *random variable* which corresponds to the position of a particle at time t. Consider observing such particles from initial time $t=$"0" to final time $t=$"1". The observation times are recorded to run from 0 to 1, for convenience. Since for each t such that $0 \leqslant t \leqslant 1$, we have associated a random variable, it is clear that we are dealing with an uncountable number of random variables.

Physicists are frequently interested in such questions as:

1. "What fraction" of particles will never rise above some level, say 10?

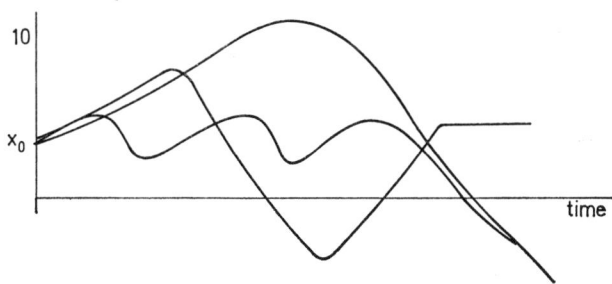

2. "What fraction" of particles will never rise above some level, say 10, or stray below some level, say -20?

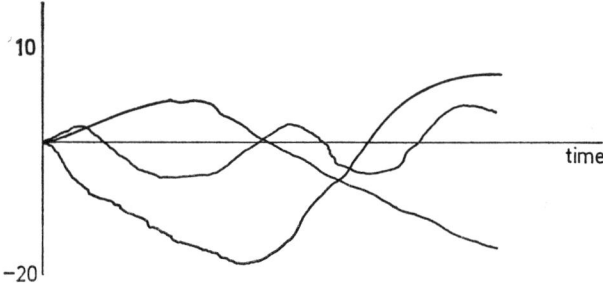

For example consider the following "paper" experiment. We are firing radiactive particles into a cylinder which has absorbing walls and a piece of fissionable material for its end.

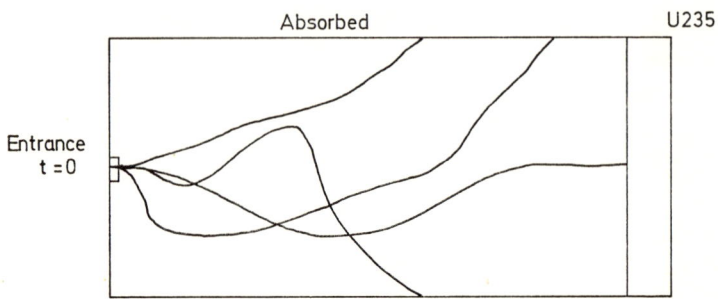

If "too few" particles reach the end, nothing will happen. If "too many" reach the end, the lab will explode. The question is "how many" particles do we fire at $t=0$ in order to get the "right number" at $t=1$?

It appears that these are questions that probability can answer. After all, the word fraction is merely the word for (Favorable Events/Total Events) which is the probability that an event can happen. The problem becomes more formidable when we realize that there are an uncountable number of random variables involved.

Let us now focus on one random variable, say $X(t)$, where t is some number between 0 and 1, and ask for the probability of $X(t)$ being between x_1 and x_2, given that at $t=0$, $X(0)=x_0$.

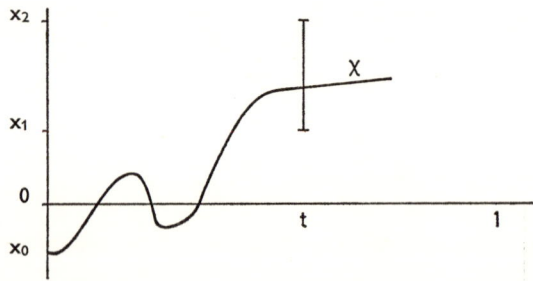

We want $\int_{x_1}^{x_2} p(x;t|x_0)$ where $p(x;t|x_0)$ is the conditional probability density of $X(t)$, given that $X(0)=x_0$.

In 1905, Albert Einstein showed that this density satisfied the partial differential equation $\partial p/\partial t = D(\partial^2 p/\partial x^2)$ where D is a certain positive physical constant. See [3]. The conditions imposed on p are

(a) $p \geq 0$

(b) $\int_{-\infty}^{\infty} p(x; t|x_0) \, dx = 1$

(c) $\lim_{t \to 0+} p(x; t|x_0) = 0$ for $x \neq x_0$.

The unique solution to this system is

$$p(x; t|x_0) = \frac{1}{\sqrt{2\pi \, 2Dt}} \exp\left[\frac{-(x-x_0)^2}{4Dt}\right].$$

This is a normal density with mean x_0, variance $2Dt$. It is also true that any finite collection of random variables $x(t_1)$, $x(t_2)$, ..., $x(t_n)$, $0 < t_1 < t_2 ... < t_n \leq 1$ has a joint normal distribution. This means that to compute $P\{-5 \leq x(t_i) \leq 5, t_i = \frac{1}{4}, \frac{1}{2}, \frac{3}{4}\}$ we would only have to compute

$$\int_{-5}^{5} \int_{-5}^{5} \int_{-5}^{5} f(x_1, x_2, x_3; t_1, t_2, t_3) \, dx_1 \, dx_2 \, dx_3$$

where f is a certain joint normal distribution. But if you think back to questions the physicist asked, this is not of much help. His question was more of the type: What is $P\{-5 \leq x(t) \leq 5, 0 \leq t \leq 1\}$? Let us try to proceed. Let $0 < t_1 < t_2 < t_3 < ... < t_n \leq 1$ where we will think of $n \to \infty$ and $\|\Delta\| = \max(t_1 - 0, t_2 - t_1, ..., t_n - t_{n-1}) \to 0$. Then, heuristically,

$P\{-5 \leq x(t) \leq 5, 0 \leq t \leq 1\}$

$$= \lim_{\substack{n \to \infty \\ \|\Delta\| \to 0}} \int_{-5}^{5} (n) ... \int_{-5}^{5} f(x_1, x_2, ..., x_n; t_1, t_2, ..., t_n) \, dx_1 \, dx_2 ... \, dx_n$$

If you have ever calculated a joint normal distribution for even $n = 2$, you would now throw up your hands, quit, and walk out!

But luckily, it is possible to show that

$$P\{-\alpha \leq x(t) \leq \alpha, 0 \leq t \leq 1\} = \frac{4}{\pi} \sum_{k=0}^{\infty} \frac{(-1)^k}{2k+1} \exp\left[-\frac{(2k+1)^2 \pi^2}{16\alpha^2}\right] \quad \text{for } D = 1.$$

With $\alpha = 5$, we have

$$= 4/\pi \{e^{-(\pi^2/400)} - 1/3 \, e^{-(9\pi^2/400)} + 1/5 \, e^{-(25\pi^2/400)} - 1/7 \, e^{-(49\pi^2/400)}\}$$
$$= .99 \text{ with error less than } 1/9 \, e^{-(81\pi^2/400)}.$$

If $\alpha = 1$, then $p = .68$ with error $\leq .01$ (one term).

This tells us that 99% of the particles fired will reach the end. If it had proved more convenient to have had a larger cylinder, we again could have calculated the % of particles reaching the end, with $\alpha=10$, etc.; it would be more than 99%, requiring less particles at entry.

Such is the type of question which is being studied now in the subject of stochastic processes. A stochastic process is a collection of random variables $\{x(t), t \in S\}$ where it is usually assumed that S has an infinite number of elements. Stochastic processes arose from the study of certain problems in physics. But they are also being applied in statistics, engineering, insurance, biology etc. ...

1.2. A Tobacco Industry Example

As a second example, consider a problem the cigarette industry is currently facing. Let us assume that the tobacco industry claims that residual tar in a carton of cigarettes (after filtering) is normally distributed with mean 10 (units) and variance 1 (unit). The U.S. Health Service wishes to test this hypothesis, with principal emphasis on the mean. Let X_i = the amount of residual tar in the ith carton. Let $Y_i = X_i - 10$, the excess or the deficiency from the mean. The Health Service proposes to test one carton at a time, up to a maximum of 500 cartons, but will stop when the first $Z_n = (1/\sqrt{500}) \sum_{i=1}^{n} Y_i$ exceeds some number. What should this number be so that the Health Service can have 99% confidence in its conclusion?

We are asking: What number k should be chosen such that $P[\text{maximum}_{n=1, 2, \ldots, 500} Z_n \geq k] = .01$? This says there is only a 1% chance of a fluke excess above k. Let us graph a typical collection of ordered pairs $(1/500, Z_1)$, $(2/500, Z_2)$, $(3/500, Z_3)$, etc.

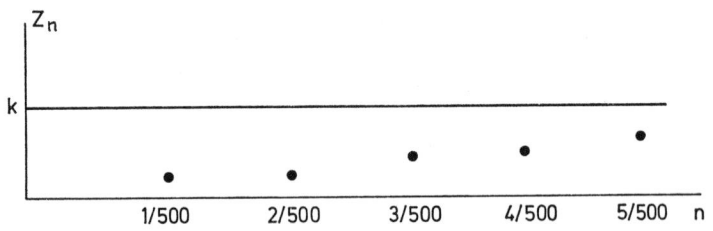

The factor $1/\sqrt{500}$ was introduced to aid in graphing the sequence $\{Z_n\}$ in a compact fashion.

Several facts now suggest that we approximate the desired probability by an analogous probability for Brownian motion. The preceding graph suggests an approximation by a continuous curve. Under the null hypothesis, Z_n has a normal distribution with $E(Z_n) = 0$, $\text{Var}(Z_n) = n/500$.

For the Brownian motion process $\{x(t), 0 \leq t \leq 1\}$, $x(t)$ has a normal distribution with $E\{X(t)\}=0$, and Var $[X(t)]=t$. Furthermore, $x(0)=0$ and we can quite plausibly make $Z_0=0$.

Hence a reasonable approximation is

$$P[\underset{n=1,2,\ldots,500}{\text{maximum}} Z_n \geq k] \doteq P[\sup_{0 \leq t \leq 1} x(t) \geq k]$$

where the latter refers to the Brownian motion process. Now it can be shown that

$$P[\sup_{0 \leq t \leq 1} x(t) \geq k] = 2P[x(1) \geq k]$$

$$= \frac{\sqrt{2}}{\sqrt{\pi}} \int_k^\infty e^{-y^2/2} dy,$$

and the k which makes that equal .01 is $k=2.575$. This solves the problem. If some $Z_n \geq 2.575$, the Health Service says the hypothesis is unrealistic; if all $Z_n < 2.575$, the hypothesis is accepted.

1.3. Noise calculations

Assume that you have some system which is sending a message in time. Either unintentionally, or intentionally (as in war-time) there is a noise present. We wish to filter out the noise and obtain the true signal.

A typical example of noise is the shot effect in vacuum tubes. This is due to random fluctuations in the intensity of the stream of electrons flowing from the cathode to the anode. To express this as a stochastic process, assume that the arrival of an electron at the anode at time $t=0$ produces an effect $E(t)$ at some point t in the output circuit. Let us also assume that the output circuit is such that the effects of the various electrons add linearly. We may then represent the total effect at time t due to all the electrons as

$$X(t) = \sum_{\text{all } t_k < t} E(t - t_k)$$

where the kth electron arrives at t_k, and the series is assumed to converge. Assume that the electrons arrive independently. Then it is plausible that $X(t)$ is asymptotically a normal variable, and $\{X(t), 0 \leq t \leq 1\}$ is a normal stochastic process.

Perhaps a word should be said about "negative" noise. Assume we are transmitting a signal at a certain frequency. If the filter which is passing this frequency allows other frequencies to pass through it, noise is present.

Frequencies above the desired frequency are positive noise; those below are negative noise.

N. Wiener in [11] suggested that an appropriate model for some noise was the Brownian motion process. If we let $X(t) = $ noise at time t, then

$$P[\max_{0 \leqslant t \leqslant 1} x(t) \leqslant \alpha] = \begin{cases} \sqrt{\dfrac{1}{\pi}} \int_0^\alpha e^{-u^2/2} du, & \alpha \geqslant 0 \\ 0, & \alpha < 0 \end{cases}$$

and

$$P[\max_{0 \leqslant u \leqslant 1} |x(u)| \leqslant \alpha] = \frac{4}{\pi} \sum_{k=0}^{\infty} \frac{(-1)^k}{2k+1} \exp\left\{-\frac{(2k+1)^2 \pi^2}{8\alpha^2}\right\}, \quad \alpha > 0.$$

These results give some measure of the probability of noise staying within certain bounds.

The current trend is to use another model for *white* noise, called the Ornstein-Uhlenbeck process. This process will be discussed in a later chapter. For references using this model, the reader is referred to pages 96, 97, and 113 of [9] by E. Parzen, and to [4] and [5] by Takeyuki Hida. The latter reference not only discusses Gaussian white noise but Poisson white noise which might provide further application of Chapter 3.

A very interesting book on this subject is *Filtering for Stochastic Processes with Applications to Guidance* by R. S. Bucy and P. D. Joseph, Interscience, New York, 1968. It gives a detailed derivation of the Kalman-Bucy filter. Quoting from page viii of the preface: "Basically, the Kalman-Bucy filter may be viewed as an algorithm which sequentially computes in real time the conditional distribution of the signal process given the observation process."

There are several journals which contain articles modeling noise or the signal with Gaussian-Markov processes. Among the papers in the *SIAM Journal on Control*, we would mention the paper "On the Differential Equations Satisfied by Conditional Probability Densities of Markov Processes, with Applications," by H. J. Kushner, which appeared on pages 106–119 of Volume 2 (1965). Among the papers in the *International Journal of Control*, we give reference to "Optimal Control of Markov Stochastic Systems Which Have Random Variation of Gain of Plant," by N. G. F. Sancho, which appeared on pages 487–496 of Volume 3 (1966), and to "Optimal Filtering for Gauss-Markov Noise," by E. B. Stear and A. R. Stubberud, which appeared on pages 123–130 of Volume 8 (1968).

The following problem in noise will be considered in Chapter 4. The description is based on page 32 of a paper by Darling and Siegert [2]. Consider "the problem of finding the probability distribution of the noise output of a radio receiver consisting of a linear amplifier, an arbitrary detector, and a second linear amplifier. Let $x(\tau)$ be the output voltage of the first amplifier at time $\tau \geqslant 0$ before observation (it is convenient to choose the time scale positive into the past), $\varphi[x(\tau)]$ the output voltage of the detector at the same time, and $K(\tau)$ the output of the second amplifier at the time of observation if a δ-function pulse is applied to it at the time τ. The output voltage V of the second amplifier in response to $x(\tau)$ is then $V = \int_0^t K(\tau)\varphi[x(\tau)]d\tau$ if the noise was turned on at a time $t \geqslant 0$ before observation. If the input of the first amplifier is white noise, $x(\tau)$ is a Gaussian random function "...." If the first amplifier is equivalent to a network with lumped circuit elements and its input is white noise, $x(\tau)$ is also a component of a Markoff process." The authors of [2] mention that in the case $\varphi(x) \equiv x^2$, the distribution problem is reduced to the solution of an integral equation. Chapter 4 presents a different approach to this distribution problem relying on another paper co-authored by Siegert.

Exercises

1. In the controlled fission section, verify that the conditional probability density satisfies the partial differential equation and conditions a), b), and c).
2. Compute $P\{-1 \leqslant x(t) \leqslant 1, \ 0 \leqslant t \leqslant 1\}$ for the Brownian motion process so that the error is $\leqslant .001$. How many terms did it take?
3. Compute $P[\sup_{0 \leqslant t \leqslant 1} x(t) \geqslant 1]$ for the Brownian motion process.

REFERENCES

1. BACHELIER, L. (1900). Théorie de la Speculation. *Ann. Sci. École Norm Sup. III*, 21–86.
2. DARLING, D. A. and SIEGERT, A. J. F. (1957). A systematic approach to a class of problems in the theory of noise and other random phenomena, Part I. *IRE Trans. Information Theory* **3**, 32–37.
3. EINSTEIN, A. (1905; 1956). *Investigations on the Theory of the Brownian Movement*, edited with notes by R. Fürth, translated by A. D. Cowper. Dover, New York.
4. HIDA, T. (1967). Finite dimensional approximations to white noise and brownian motion. *J. Math. Mech.* **16**, 859–866.
5. — (1970). *Stationary Stochastic Processes*. Princeton Univ. Press, Princeton.
6. HOEL, P. G. (1971). *Introduction to Mathematical Statistics*, 4th Edition. John Wiley and Sons, New York.
7. KAC, M. (1969). Some mathematical models in science. *Science* **166**, 695–699.
8. LUNDBERG, F. (1903). *Approximerad framställning av sannolikhetsfunktionen. Återförsäkring av kollektivrisker*. Diss. Almqvist & Wiksell, Uppsala.
9. PARZEN, E. (1962). *Stochastic Processes*. Holden-Day, San Francisco.
10. WIENER, N. (1923). Differential space. *J. Math. Phys. II*, 131–174.
11. — (1949). *Time Series*. M.I.T. Press, Cambridge.

CHAPTER 2

SOME MATHEMATICAL PRELIMINARIES

2.0 Summary

In this chapter, we will discuss Stieltjes integrals, and Laplace transforms. It is felt that the treatment of Stieltjes integrals is a good preview of certain features of Chapters 3 and 4.

2.1. Historical remarks on integration theory

There are certain dates in the theory of integration which should be mentioned. The calculus which we all start with was invented by Sir Isaac Newton (1642–1727) and Gottfried Wilhelm Leibniz (1646–1716). G. F. B. Riemann (1826–1866) contributed much to this form of calculus and we frequently refer to the usual integral as the Riemann integral. T. J. Stieltjes (1856–1894) in 1894 introduced a generalization of the Riemann integral. Stieltjes integrals proved useful in physics by allowing the treatment of continuous and discrete phenomena within one theoretical framework. Essentially this describes what we will gain by using Stieltjes integrals in our probabilistic considerations. Henri Lebesgue (1875–1943) devised a more abstract integral which is very useful in probability theory. Norbert Wiener (1894–1964) developed an integral over a space of continuous functions. Wiener and Lebesgue integrals will be considered in Chapter 4.

A presentation of any one of these integrals would follow a certain pattern, with four subsections. First, the integral would be defined. Secondly, conditions would be established which are sufficient for the integral to exist. The third stage is to establish properties of the integral. Finally, a method of calculation is established which is easier than that of the definition.

2.2. Definition of a Stieltjes integral

Let Δ be a subdivision of $[a, b]$, i.e., Δ is a set of numbers $\{x_0, x_1, x_2, ..., x_n\}$ with $a = x_0 < x_1 < x_2 < ... < x_n = b$. The norm of Δ is denoted and defined by $\|\Delta\| = \max(x_1-x_0, x_2-x_1, ..., x_n-x_{n-1})$. The Stieltjes integral of $f(x)$ with respect to $\alpha(x)$ from a to b is denoted and defined by

$$\int_a^b f(x)\,d\alpha(x) = \lim_{\|\Delta\|\to 0} \sum_{k=1}^n f(\zeta_k)[\alpha(x_k) - \alpha(x_{k-1})]$$

where $x_{k-1} \leqslant \zeta_k \leqslant x_k$, $k=1, 2, ..., n$.

Note that with $\alpha(x) = x$ this reduces to the Riemann integral.

As an example, let $f(x) = x^2$, $0 \leqslant x \leqslant 2$, and $\alpha(x) = \begin{cases} 0, 0 \leqslant x < 1 \\ 1, 1 \leqslant x \leqslant 2 \end{cases}$. Then $\int_0^2 f(x)\,d\alpha(x) = 1$.

2.3. Existence of the Integral

The Stieltjes integral may fail to exist if one does not impose some conditions on $f(x)$ and $\alpha(x)$. Thus, if $f(x) = \begin{cases} 1, 0 \leqslant x < 1 \\ 2, 1 \leqslant x \leqslant 2 \end{cases}$ and $\alpha(x) = \begin{cases} 5, 0 \leqslant x < 1 \\ 6, 1 \leqslant x \leqslant 2 \end{cases}$ the limit involved in the definition of the Stieltjes integral of $f(x)$ with respect to $\alpha(x)$ will not exist.

THEOREM 1. *If $f(x)$ is continuous on $[a, b]$, and $\alpha(x)$ is nondecreasing on $[a, b]$, then $\int_a^b f(x)\,d\alpha(x)$ exists.*

The proof of this may be found in several textbooks, for example on pages 176–178 of Widder [3].

COROLLARY. *The theorem holds if $\alpha(x)$ is nonincreasing on $[a, b]$.*

THEOREM 2. *If $f(x)$ is continuous on $[a, b]$, and $\alpha(x) = \alpha_1(x) + \alpha_2(x)$ where $\alpha_1(x)$ is nondecreasing on $[a, b]$ and $\alpha_2(x)$ is nonincreasing on $[a, b]$, then $\int_a^b f(x)\,d\alpha(x)$ exists and equals $\int_a^b f(x)\,d\alpha_1(x) + \int_a^b f(x)\,d\alpha_2(x)$.*

This holds by routine application of the theorem about the limit of a sum of two variables.

2.4. Properties of Stieltjes Integrals

The Stieltjes integral enjoys many of the properties of the Riemann integral. For example, the Stieltjes integral of the sum of two functions is the sum of the appropriate Stieltjes integrals, provided they exist. The Stieltjes integral over a range $[a, b]$ may be split into two Stieltjes integrals over $[a, c]$ and $[c, b]$ where $a < c < b$. Although many advanced calculus books recite these properties, they are summarized very nicely on page 155 of Widder [3], and that is the form in which we will present them.

We assume that k is a constant, and the functions $f(x)$ and $\alpha(x)$, with or without subscripts, are, respectively, continuous and nondecreasing on $[a, b]$.

I. $\displaystyle\int_a^b d\alpha(x) = \alpha(b) - \alpha(a).$

II. $\displaystyle\int_a^b f(x)\,d[\alpha(x) + k] = \int_a^b f(x)\,d\alpha(x).$

III. $\displaystyle\int_a^b kf(x)\,d\alpha(x) = k\int_a^b f(x)\,d\alpha(x).$

IV. $\displaystyle\int_a^b [f_1(x) + f_2(x)]\,d\alpha(x) = \int_a^b f_1(x)\,d\alpha(x) + \int_a^b f_2(x)\,d\alpha(x).$

V. $\displaystyle\int_a^b f(x)\,d[\alpha_1(x) + \alpha_2(x)] = \int_a^b f(x)\,d\alpha_1(x) + \int_a^b f(x)\,d\alpha_2(x).$

VI. $\displaystyle\int_a^b f(x)\,d\alpha(x) = \int_a^c f(x)\,d\alpha(x) + \int_c^b f(x)\,d\alpha(x),\ a < c < b.$

VII. If $f_1(x) \leqslant f_2(x)$, for $a \leqslant x \leqslant b$, then
$$\int_a^b f_1(x)\,d\alpha(x) \leqslant \int_a^b f_2(x)\,d\alpha(x).$$

VIII. $\left|\displaystyle\int_a^b f(x)\,d\alpha(x)\right| \leqslant \int_a^b |f(x)|\,d\alpha(x).$

IX. $\left|\displaystyle\int_a^b f(x)\,d\alpha(x)\right| \leqslant [\alpha(b) - \alpha(a)] \max_{a \leqslant x \leqslant b} |f(x)|.$

The proofs for most of these properties are rather straightforward. One expresses the Stieltjes integral on the left side of each property in terms of its limit, and then uses the various theorems about limits to achieve the right side of the property. The reader would profit by doing several of these proofs.

2.5. Easy methods of calculation

Obviously, few Stieltjes integrals would be calculated if the only calculation device was the definition. Luckily, the most important types of Stieltjes integrals can be calculated by the following methods.

THEOREM 1. Let $\alpha(x)$ be a step-function with jumps of amounts h_j at the points c_i where $a < c_1 < c_2 < ... < c_n < b$. Let $f(x)$ be continuous on $[a,b]$. Then

$$\int_a^b f(x)\,d\alpha(x) = \sum_{k=1}^n h_k f(c_k).$$

Proof. Consider a set of numbers $\{d_0, d_1, ..., d_n\}$ with the property that

$$d_0 = a < c_1 < d_1 < c_2 < d_2 < ... < d_{n-1} < c_n < b = d_n.$$

By a mild extension of property VI,

$$\int_a^b f(x)\,d\alpha(x) = \sum_{i=0}^{n-1} \int_{d_i}^{d_{i+1}} f(x)\,d\alpha(x).$$

Now for any fixed i,

$$\int_{d_i}^{d_{i+1}} f(x)\,d\alpha(x) = f(c_{i+1})[\alpha(d_{i+1}) - \alpha(d_i)] = f(c_{i+1})h_{i+1}.$$

Letting $k = i+1$, the result follows.

This theorem allows one to express convergent series as Stieltjes integrals. For example, assume that $\sum_{n=1}^\infty a_n < \infty$. Define a polygonal function in this manner:

$$a_x = \begin{cases} a_1, & 0 \leq x < 1 \\ a_n, & x = n \text{ for } n = 1, 2, 3, ... \\ a_n + (x-n)(a_{n+1} - a_n) & \text{for } n < x < n+1 \text{ and } n = 1, 2, 3, ... \end{cases}$$

If $[x]$ represents the greatest integer function, then $\sum_{n=1}^\infty a_n = \int_0^\infty a_x\,d[x]$.

The second easy method of calculating Stieltjes integrals is the following.

THEOREM 2. Assume that $f(x)$ is continuous on $[a,b]$, and that $a(x)$ and $a'(x)$ are both continuous on $[a,b]$. Then

$$\int_a^b f(x)\,d\alpha(x) = \int_a^b f(x)\,\alpha'(x)\,dx.$$

Note that the right hand integral is an ordinary Riemann integral. The proof of this theorem may be found on pages 156 and 173–174 of [3]. For example, if $f(x) = x^2$ and $\alpha(x) = \sin x$ for $0 \leq x \leq \pi$, then

$$\int_0^\pi f(x)\,d\alpha(x) = \int_0^\pi x^2 \cos x\,dx.$$

Exercises

1. Evaluate $\int_0^1 x^3 \, dx^2$

2. Let $g(x) = \begin{cases} 0, & x < 1 \\ \frac{1}{3}, & 1 \leq x < 2 \\ \frac{2}{3}, & 2 \leq x < 3 \\ 1, & 3 \leq x \end{cases}$

   Compute $\int_0^5 x^2 \, dg(x)$

3. Evaluate $\int_0^\pi (x+2) \, d \sin x$

4. Evaluate $\int_0^{2\pi} \sin x \, d \cos x$

5. For $g(x)$ as in #2, compute $\int_0^\pi \sin x \, dg(x)$.

6. If $\alpha(x) = 7$ except in the interval $(-2, 2)$ where $\alpha(x) = x$, find $\int_{-3}^8 x^4 \, d\alpha(x)$.

7. For any continuous function $f(x)$ on $0 \leq x \leq 20$, define $a(x)$ so that
$$\int_0^{20} f(x) \, d\alpha(x) = f(0) - f(2) + 2f(10) - 7.1f(7.1) + 9f(20).$$

The reader will recall from ordinary calculus a technique called integration by parts. Surprisingly, that symbology is really borrowed from the comparable theorem for Stieltjes integrals.

THEOREM. Assume that $f(x)$ is continuous on $[a, b]$ and $\alpha(x)$ is nondecreasing on $[a, b]$. Then $\int_a^b f(x) \, d\alpha(x) = -\int_a^b \alpha(x) \, df(x) + \alpha(b) f(b) - \alpha(a) f(a)$.

The proof of this will be found on page 160 of [3]. It may also be extended to allow $\alpha(x)$ to be the sum of a nondecreasing function and a nonincreasing function on $[a, b]$.

8. Rework #3 above, using integration by parts.

2.6. Applications to Probability

We are now in a position to apply these ideas to probability definitions, theorems, and problems.

We will begin by giving a formal definition of a random variable. The domain of definition of a random variable, say X, is a sample space, which we will denote by Ω. A sample point in Ω will be represented by ω. A subset A of Ω will be called an *event*. A *family* F of events is a collection of subsets of Ω with the following properties:

(1) $\Omega \in F$.

(2) If $A \in F$, then $A^C \in F$.

(3) If $A_1, A_2, \ldots$ belong to F, then $\bigcup_{i=1}^{\infty} A_i \in F$.

We assume that a *probability function* $P(\cdot)$ has been defined on F with the properties:

(1) $P(A) \geq 0$ for $A \in F$.

(2) $P(\Omega) = 1$.

(3) If $A_1, A_2, \ldots$ belong to F, and $A_j \cap A_k = \emptyset$ for $j \neq k$, then

$$P\left[\bigcup_{i=1}^{\infty} A_i\right] = \sum_{i=1}^{\infty} P(A_i)$$

Let R be the real line. The family β of *Borel* sets of R is the smallest *family* of sets of real numbers that also contains all open, semiopen, and closed intervals.

A random variable X has the property that for every Borel set B of real numbers, the set $\{\omega: X(\omega) \in B\} \in F$.

We draw these ideas together in the:

Definition. *A random variable* X *is a real finite valued function defined on a sample space* Ω *on whose family* F *of events a probability function* $P(\cdot)$ *has been defined, and has the property that for every Borel set* B *of real numbers, the set* $\{\omega: X(\omega) \in B\} \in F$.

Let us recall that the distribution function of a real-valued random variable X is a function $F(x)$ with the properties:

(1) $F(x)$ is nondecreasing, $-\infty < x < \infty$;

(2) $F(x+) = F(x)$, i.e. $F(x)$ is continuous from the right for $-\infty < x < \infty$;

(3) $\lim_{x \to -\infty} F(x) = 0$;

(4) $\lim_{x \to +\infty} F(x) = 1$.

Note that

$$\int_{-\infty}^{\infty} dF(x) = \lim_{\substack{A \to +\infty \\ B \to -\infty}} \int_B^A dF(x)$$

$$= \lim_{\substack{A \to +\infty \\ B \to -\infty}} [F(A) - F(B)], \text{ by Property I}$$

$$= 1 - 0 = 1.$$

With few exceptions, any distribution function $F(x)$ can be represented as $F(x) = a_1 F_1(x) + a_2 F_2(x)$ where a_1 and a_2 are non-negative numbers such that $a_1 + a_2 = 1$, $F_1(x)$ and $F_2(x)$ are distribution functions such that

$$F_1(x) = \int_{-\infty}^x F_1'(t) \, dt,$$

and $\qquad F_2(x) = $ a step function.

When $a_2 = 0$, $F(x)$ may be thought of as the distribution function for a continuous random variable, whereas when $a_1 = 0$, we have a discrete random variable.

DEFINITION. Let X be a random variable with distribution function $F(x)$, $-\infty < x < \infty$. The kth moment of X is

$$E[X^k] = \int_{-\infty}^{\infty} x^k dF(x).$$

The first moment is referred to as the mean value. The variance of X is defined to be

$$\text{Var}(X) = E\{[X - E(X)]^2\} = \int_{-\infty}^{\infty} [y - E(X)]^2 \, dF(y),$$

and the standard deviation of X is

$$\sqrt{\text{Var}(x)}.$$

THEOREM. $\text{Var}(X) = E(X^2) - [E(X)]^2$

Proof. $\operatorname{Var}(X) = \int_{-\infty}^{\infty} [y - E(X)]^2 \, dF(y)$

$$= \int_{-\infty}^{\infty} [y^2 - 2y \, E(X) + (E(X))^2] \, dF(y)$$

$$= \int_{-\infty}^{\infty} y^2 \, dF(y) - 2E(X) \int_{-\infty}^{\infty} y \, dF(y) + (E(X))^2 \int_{-\infty}^{\infty} dF(y)$$

$$= E(X^2) - (E(X))^2.$$

Theorem. If $F(x) = \begin{cases} 0, & x < k \\ 1, & x \geq k \end{cases}$, then $E(X) = k$ and $\operatorname{Var}(X) = 0$.

Proof. $E(X^i) = \int_{-\infty}^{\infty} x^i \, dF(x) = k^i.$

$$\operatorname{Var}(X) = E(X^2) - (E(X))^2$$
$$= k^2 - k^2 = 0.$$

If X is a random variable with distribution function $F(x)$, $-\infty < x < \infty$, then for a very large class of functions g,

$$E[g(X)] = \int_{-\infty}^{\infty} g(x) \, dF(x).$$

Theorem: $E(aX + b) = aE(X) + b$, assuming that the moments are finite.

Proof: $E(aX + b) = \int_{-\infty}^{\infty} (ax + b) \, dF(x)$

$$= a \int_{-\infty}^{\infty} x \, dF(x) + b \int_{-\infty}^{\infty} dF(x)$$

$$= aE(X) + b.$$

Theorem: $\operatorname{Var}(aX + b) = a^2 \operatorname{Var}(X).$

Proof: $\operatorname{Var}(aX + b) = E\{[aX + b - E(aX + b)]^2\}$

$$= E\{[aX + b - aE(X) - b]^2\}$$
$$= E\{a^2[X - E(X)]^2\}$$
$$= \int_{-\infty}^{\infty} a^2 (X - E(X))^2 \, dF(x)$$

$$= a^2 \int_{-\infty}^{\infty} (x - E(X))^2 \, dF(x)$$

$$= a^2 \operatorname{Var}(X).$$

DEFINITION: If a random variable X has mean $E(X)$ and variance $\operatorname{Var}(X)$, its standardized version is

$$\frac{X - E(X)}{\sqrt{\operatorname{Var}(X)}}.$$

THEOREM. If Z denotes the standardized version of X, $E(Z) = 0$, and $\operatorname{Var}(Z) = 1$.

Proof. Apply the previous two theorems.

Exercises

9. Assume that $F(x) = \begin{cases} 0, & x < 0 \\ x, & 0 \leqslant x \leqslant 1 \\ 1, & x > 1 \end{cases}$.

   Compute $E(X)$ and $\operatorname{Var}(X)$.

10. Assume that $F(x) = 0$, $x < 5$.
    $$1, \; x \geqslant 5$$

    Compute $E(X)$ and $\operatorname{Var}(X)$.

11. Assume that the random variable X has a distribution which is a weighted average of the above two distributions. That is,

    $$P[X \leqslant x] = F(x) = a_1 F_1(x) + a_2 F_2(x)$$

    where $a_1 = \tfrac{1}{4}$, $a_2 = \tfrac{3}{4}$, and

    $$F_1(x) = \begin{cases} 0, & x < 0 \\ x, & 0 \leqslant x \leqslant 1 \\ 1, & x > 1 \end{cases}$$

    and
    $$F_2(x) = \begin{cases} 0, & x < 5 \\ 1, & x \geqslant 5 \end{cases}.$$

    Compute $E(X)$ and $\operatorname{Var}(X)$.

For a random variable X, the moment generating function is defined by

$$M_X(\theta) = E[e^{\theta X}] = \int_{-\infty}^{\infty} e^{\theta y} d_y P[X \leq y],$$

provided that the integral is convergent for some θ interval including $\theta = 0$. The existence of $M_X(\theta)$ for $-c < \theta < c$ implies that derivatives of all orders exist at $\theta = 0$. Thus

$$M_X^{(k)}(\theta) = \int_{-\infty}^{\infty} y^k e^{\theta y} d_y P[X \leq y], \quad \text{and} \quad E(X^k) = M_X^{(k)}(\theta)\big|_{\theta=0}.$$

The characteristic function of a random variable X is defined by

$$C_X(\theta) = E[e^{i\theta X}] = \int_{-\infty}^{\infty} e^{i\theta y} d_y P[X \leq y].$$

For every real number y, $|e^{i\theta y}| = 1$, and thus

$$|C_X(\theta)| \leq \int_{-\infty}^{\infty} |e^{i\theta y}| d_y P[X \leq y] = \int_{-\infty}^{\infty} d_y P[X \leq y] = 1.$$

Thus the characteristic function always exists, which is not true of the moment generating function. Provided that $E(X)$ and $E(X^2)$ exist,

$$E(X) = i^{-1} \frac{dC_X(\theta)}{d\theta}\bigg|_{\theta=0} \quad \text{and} \quad E(X^2) = -\frac{d^2 C_X(\theta)}{d\theta^2}\bigg|_{\theta=0}.$$

There is a one to one correspondence between distribution functions and characteristic functions. One such connecting theorem is the following:

FOURIER INVERSION THEOREM. Assume that

$$\int_{-\infty}^{\infty} |C_X(\theta)| d\theta < +\infty.$$

Then $F(y) = P[X \leq y]$ has a bounded continuous density f given by

$$f(x) = \frac{1}{2\pi} \int_{-\infty}^{\infty} e^{-i\theta x} C_X(\theta) d\theta.$$

In 1925 Paul Lévy gave the following:

LÉVY INVERSION THEOREM. If $F(x)$ is continuous for $x = x'$ and for $x = x' + h$, we have

$$F(x' + h) - F(x') = \lim_{T \to \infty} \frac{1}{2\pi} \int_{-T}^{T} \frac{1 - e^{-i\theta h}}{i\theta} e^{-i\theta x'} C_X(\theta) d\theta.$$

See [1]. Frequently it is difficult to obtain $f(x)$ or $F(x)$ from these formulas, because of the complicated contour integrals involved. However, a number of excellent numerical inversion methods have been developed recently. The reader will see two such methods in section 3.5.

2.7. Laplace and Laplace–Stieltjes transforms

Laplace transforms are similar to the moment-generating functions studied in mathematical statistics. Knowledge of the transform of a function $F(x)$ tells a great deal about $F(x)$. For the functions we will consider there is a unique correspondence between the functions and their transforms. The process of "undoing" the transform to obtain $F(x)$ is referred to as "inverting the transform." Reference [2] and many other books contain tables of functions and their Laplace transforms. Such tables can be read from either entry. That is, given a function, one can read its Laplace transform, or, given a Laplace transform, one can deduce what function produced it. This is visual inversion.

If $F(x)$ is a function of x for $x>0$, its Laplace transform is the integral

$$f(s) = \int_0^\infty e^{-sx} F(x)\, dx$$

for s a real number. For example, if $F(x)=x$, $x>0$, then $f(s)=1/s^2$, $s>0$. The process of inversion consists of finding $F(x)$ from $f(s)$. For example, if $f(s)=1/(s-A)$, $s>A$, a table of transforms reveals that $F(x)=e^{Ax}$.

Laplace transforms and their inverses enjoy the linearity property, which is most convenient. That is,

$$\int_0^\infty e^{-sx}[aF_1(x)+bF_2(x)]\, dx = af_1(s)+bf_2(s),$$

and using $I_s[f(s)]$ for the inverse Laplace transform of $f(s)$,

$$I_s[af_1(s)+bf_2(s)] = aF_1(x)+bF_2(x).$$

For example, $\quad I_s\left[a\left(\dfrac{1}{s}\right)+b\left(\dfrac{1}{s+c}\right)\right] = a+be^{-cx}.$

Laplace–Stieltjes transforms are of the form

$$f(s) = \int_0^\infty e^{-sx} dg(x).$$

Knowing $g(x)$ one can compute $f(s)$, and theoretically one can find $g(x)$ from $f(s)$. However, such tables are hard to find, and therefore it is advantageous to reduce Laplace-Stieltjes transforms to Laplace transforms. This can be done by the integration by parts formula for Stieltjes integrals.

Thus

$$\int_0^\infty e^{-sx} dg(x) = \lim_{w \to \infty} e^{-sx} g(x) \Big|_0^w + s \int_0^\infty e^{-sx} g(x)\, dx.$$

Frequently in our work

$$\lim_{w \to \infty} e^{-sw} g(w) = 0$$

and so

$$\int_0^\infty e^{-sx} dg(x) = -g(0) + s \int_0^\infty e^{-sx} g(x)\, dx.$$

Exercises

12. If $F(x) = e^{Ax}$, verify that $f(s) = 1/(s-A)$, $s > A$.
13. If $F_n(x) = x^n$, $n = 0, 1, 2, \ldots$ show that $f(s) = n!/s^{n+1}$, $s > 0$.
14. If $F(x) = \sin(Ax)$, show that $f(s) = A/(s^2 + A^2)$, $s > 0$. Use the fact that

$$\sin(Ax) = \frac{e^{iAx} - e^{-iAx}}{2i}.$$

15. If X is a non-negative valued random variable and

$$f(s) = \int_0^\infty e^{-sy} d_y P[X \leq y] \quad \text{for } s > 0,$$

derive expressions for $E(X)$ and $E(X^2)$ involving derivatives of $f(s)$.

For a non-negative valued random variable, the moment generating function is

$$M_X(\theta) = E[e^{\theta X}] = \int_0^\infty e^{\theta y} d_y P[X \leq y].$$

Thus the process of obtaining moments from $M_X(\theta)$ is almost identical to the process using $f(s)$.

16. A certain random variable Z in collective risk theory has been found to produce $f(s) = s\lambda \{s(1+\lambda) - 1 + (1+s)^{-1}\}^{-1}$, for $s > 0$, $\lambda > 0$. Compute the mean and variance for Z.

For a bivariate normal density, $f(x, y)$, there is a moment generating function involving two parameters:

$$M(\theta_1, \theta_2) = \int_{-\infty}^{\infty} \int_{-\infty}^{\infty} e^{\theta_1 x + \theta_2 y} f(x, y) \, dx \, dy.$$

Theorem 1 of the next chapter is analogous, although it uses Laplace and Laplace–Stieltjes transforms. Its double Laplace transform is eventually used to obtain moments.

REFERENCES

1. Lévy, P. (1925). *Calcul des Probabilités*. Gauthier-Villars, Paris.
2. SPEIGAL, M. R. (1965). *Theory and Problems of Laplace Transforms*. Schaum, New York.
3. WIDDER, D. V. (1961). *Advanced Calculus*. Prentice-Hall, Englewood Cliffs, N.J.

CHAPTER 3

THE COLLECTIVE RISK STOCHASTIC PROCESS

3.0. AN OVERVIEW OF THE SUBJECT

There are at least three good reasons for studying collective risk theory: (1) to learn about a valuable tool for insurance management; (2) to enjoy the intellectual beauty of the subject; and (3) to learn of its mathematical methods in order to apply them outside of insurance.

Within this chapter will be found examples on the use of collective risk theory to compute net stop-loss reinsurance premiums, to set retention limits for a type of insurance, to set the margin above the net premiums so necessary to keep the company solvent, and to determine the amount of initial capital to allocate to a new line of business.

For those who see beauty in mathematics, both pure and applied, much pleasure can be derived by studying the evolution of the subject within the many papers cited here and elsewhere.

Although this chapter is partially phrased within the context of insurance, many of these techniques can be applied to probability problems which arise in the theory of queues, dams, storage, order statistics, physics, astronomy, and elsewhere.

Let $X_1, X_2, \ldots$ be a sequence of independent, identically distributed random variables. They will correspond to insurance claims. The number of insurance claims in a fixed time period is a random variable. We will now discuss its distribution based on page 19 of [21]. It is clear that as the company grows the number of claims expected in a similar fixed time period will be larger. We will assume that the numbers of claims in any two disjoint intervals on the time scale are independent random variables. Let τ be an arbitrary point on the natural time scale, and consider the probability that there will be just one claim during the natural time interval τ to $\tau + \Delta\tau$. Since we have assumed atht this is independent of previous claims, it is natural to assume that this probability is proportional to $\Delta\tau$ for *small values of* $\Delta\tau$. Thus

$$\text{Prob [one claim in } [\tau, \tau + \Delta\tau]] = \lambda_\tau \Delta\tau + o(\Delta\tau)$$

where $o(\Delta\tau)$ denotes a quantity which becomes small compared with $\Delta\tau$ as $\Delta\tau \to 0$. Perhaps $\lambda_\tau = 200(1-e^{-2\tau})$ would be appropriate for a growing company. In any event, we assume λ_τ is a bounded and measurable function of τ for all $\tau \geqslant 0$. We also assume that the probability of more than one claim during the interval $(\tau, \tau+\Delta\tau)$ is of the form $o(\Delta\tau)$.

We now introeuce an operational time t, related to the natural time τ by the equation

$$t = \int_0^\tau \lambda_u \, du.$$

Thus, for our example,

$$t = 200[\tau + \tfrac{1}{2} e^{-2\tau} - \tfrac{1}{2}].$$

Hence $\tau=0$ implies that $t=0$, whereas, for $\tau \geqslant 5$, $t \doteqdot 200\tau - 100$. In fact, if 5 calendar periods have elapsed and we measure anew, $\lambda_u \doteqdot 200$ and $t \doteqdot 200\tau$.

Therefore, $dt = \lambda_\tau d\tau$ and the probability of just one claim during the interval $(t, t+\Delta t)$ will be $\Delta t + o(\Delta t)$, while the probability of more than one claim during the same interval is $o(\Delta t)$.

If we now let $N(t)$ be the random number of claims up through operational time t, our hypotheses dictate that $\{N(t), t \geqslant 0\}$ is a Poisson stochastic process. This is proved in a number of text-books on probability and statistics. Hence

$$P[N(t+s) - N(t) = k] = \frac{e^{-s} s^k}{k!},$$

$k=0, 1, 2, \ldots$ for $s>0$. With $N(0)=0$, it is clear that $E[N(t)]=t$. Thus, operational time allows us to measure time by the expected number of claims during the period. We assume that the $N(t)$ process is independent of the X_i's and hence any collection $N(t_1), N(t_2), \ldots, N(t_n)$ is independent of any finite collection $X_{i_1}, X_{i_2}, \ldots, X_{i_k}$.

In our example, if we start our observations at $\tau=5$ calendar periods, then $t=200$ would correspond to one calendar period and $\sum_{i=1}^{N(200)} X_i$ would be the total claims in one calendar period. It is a random number of random variables. The reader may wish to consult part of Chapter 12 of Feller's book, [22]. Insurance companies are interested in

$$P\left[\sum_{i=1}^{N(T)} X_i \leqslant \alpha\right] = F(\alpha, T),$$

the distribution of total claims through time T. The same type problems arise in queueing theory, water control theory and elsewhere. In water control, the X_i's represent the random amounts of inflow, $N(t)$ represents the random number of inputs through time t, and $\sum_{i=1}^{N(t)} X_i$ would be the aggregate input of water through time t at one point of space.

If $p_1 = E(X)$, then the insurance company would charge $p_1 t = p_1 E[N(t)]$ as the aggregate net risk premium. In water control, $p_1 t$ would be the aggregate water released through the dam through time t. They would also add a security loading of λt. If the company starts with a risk reserve of size u, it is of interest to know

$$P\left[\max_{0 \leqslant t \leqslant T} \left(\sum_{i=1}^{N(t)} X_i - (p_1 + \lambda)t\right) \leqslant u\right],$$

the probability of not going broke through time T, and

$$P\left[\max_{0 \leqslant t < \infty} \left(\sum_{i=1}^{N(t)} X_i - (p_1 + \lambda)t\right) \leqslant u\right],$$

the probability of avoiding eventual ruin. Conventionally, the probability of eventual ruin is denoted by $\psi(u)$. The above probability is $1 - \psi(u)$ which we will label $\psi^*(u)$. If $A =$ initial content of the dam,

$$A + \max_{0 \leqslant t \leqslant T} \left\{\sum_{i=1}^{N(t)} X_i - (p_1 + \lambda)t\right\}$$

is the maximum content of the dam through time T.

$$\left\{x(t) = \sum_{i=1}^{N(t)} X_i, 0 \leqslant t \leqslant T\right\}$$

is a stochastic process.

The basic sample space for a stochastic process is a set of functions. Each such function represents one possible realization of the process through time. On occasion, the brief notation $X(t)$ will be enlarged to $X(t, w)$, where w is a sample function.

For a sample function w, we have the graph of $x(t, w)$:

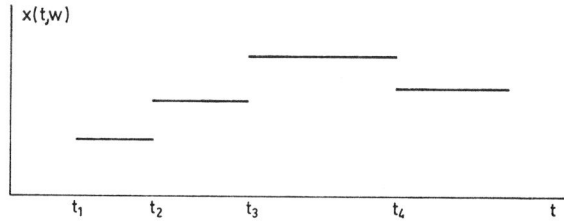

The positive jumps equal death claims, for example. The negative jumps equal negative claims: for example, an annuitant dies and a reserve is released. If each X_i could have assumed only the values 1 or 0, it would have been called the Poisson process. With more general X_i's it is called the compound Poisson process.

$$\left\{ y(t) = \sum_{i=1}^{N(t)} X_i - (p_1 + \lambda)t, \quad 0 \leqslant t \leqslant T \right\}$$

is another stochastic process. For a sample function w, we have the graph of $y(t, w)$:

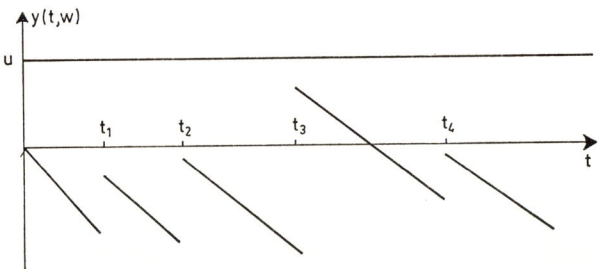

As drawn, we have positive claims at times t_1, t_2, t_3, and t_4, and a steady flow of income receipts of $(p_1+\lambda)t$ which create the negative slope line segments. We are interested in those paths that stay below u.

Both the $x(t)$ and $y(t)$ processes have stationary independent increments. An increment is a difference $y(t+s)-y(s)$, for $s \geqslant 0$, $t > 0$. To say the process has independent increments means such jumps in different time spans are independent random variables. The stationarity means, for example $P[y(20)-y(0) \leqslant \alpha] = P[y(40)-y(20) \leqslant \alpha]$.

These two properties attracted the author to a paper by Glen Baxter and Monroe Donsker [5]. It enables one to give expressions for the Laplace transforms of the distribution of total claims, and the ruin probabilities.

Let
$$\sigma(\alpha, T, k) = P\left[\max_{0 \leqslant t \leqslant T} \left\{ \sum_{i=1}^{N(t)} X_i - tk \right\} \leqslant \alpha \right].$$

Let
$$p(i\xi) = \int_{-\infty}^{\infty} e^{i\xi y} dP[X \leqslant y],$$

the characteristic function of the random variables.

THEOREM 1. (See [7].) Assume $E[|X|] = E\{|X_i|\} < \infty$. Let

$$\theta(\xi) = \int_{-\infty}^{\infty} e^{i\xi x} dP(x) - 1 - i\xi k.$$

(a) If $\theta(\xi)$ is real,

$$\int_0^\infty \int_0^\infty e^{-wT-Z\alpha} d_\alpha \sigma(\alpha, T, k) dT$$

$$= \frac{1}{w} \exp\left\{\frac{1}{2\pi} \int_w^\infty \int_{-\infty}^\infty \frac{Z}{Z^2 + \xi^2} \frac{\theta(\xi)}{s(s-\theta(\xi))} d\xi\, ds\right\}.$$

(b) If $\theta(\xi)$ is complex

$$\int_0^\infty \int_0^\infty e^{-wT-Z\alpha} d_\alpha \sigma(\alpha, T, k) dT$$

$$= \frac{1}{w} \exp\left\{\frac{1}{2\pi} \int_w^\infty \int_{-\infty}^\infty \frac{Z}{\xi(\xi - iZ)} \frac{\theta(\xi)}{s(s-\theta(\xi))} d\xi\, ds\right\}.$$

Example

Assume that $P[X_i = 1] = P[X_i = -1] = \frac{1}{2}$. Then $p(i\xi) = \frac{1}{2} e^{-i\xi} + \frac{1}{2} e^{i\xi} = \cos \xi$. If $k=0$, then $\theta(\xi) = \cos \xi - 1$. Then

$$\sigma(n, T, 0) = P\left[\max_{0 \leq t \leq T} \sum_{i=1}^{N(t)} X_i < n\right] = 1 - n \int_0^T e^{-t} \frac{I_n(t)}{t} dt$$

where $I_n(x) = i^{-n} J_n(ix)$ nad $J_n(x)$ is the Bessel function of the first kind. The author has numerically integrated this for $3 \leq n \leq 30$, $5 \leq T \leq 280$. This example can be described as coin-tossing at random times. A glance at the index of Feller [22] reveals that coin-tossing models have been extensively studied. Using formulas from Titchmarsh [42],

$$\sigma(n, T, 0) = 1 - \frac{n}{2^{n-1}(n-\frac{1}{2})(n-\frac{3}{2})\ldots(\frac{3}{2})((\frac{1}{2})(\pi)}.$$

$$\left[\int_0^T e^{-s} s^{n-1} \int_0^{\pi/2} \frac{e^{s\cos\theta} + e^{-s\cos\theta}}{2} \sin^{2n}\theta\, d\theta\, ds\right].$$

By using upper and lower approximating sums, values for $\sigma(n, T, 0)$ accurate to 2% were obtained. For $n \geq 25$, Stirling's second approximation to the gamma function ([22]) was used for part of the denominator. That is,

$$\Gamma(n+\tfrac{1}{2}) = (n-\tfrac{1}{2})(n-\tfrac{3}{2})\ldots(\tfrac{3}{2})(\tfrac{1}{2}) \pi^{\frac{1}{2}},$$

and

$$\Gamma(n+\tfrac{1}{2}) \doteq \sqrt{2\pi}(n-\tfrac{1}{2})^n e^{-(n-\frac{1}{2})+(12n-6)^{-1}}.$$

In some cases, the range of outer integration was diminished by a substitution $V = s/a$ for appropriate a's. To make a minor contribution to that literature, we will reproduce the table for

$$\sigma(n, T, 0) = P\left[\max_{0 \leq t \leq T} \sum_{i=1}^{N(t)} X_i < n\right]$$

which appears on p. 73 of [7].[1]

T	n	$\sigma(n, T, 0)$	T	n	$\sigma(n, T, 0)$
5	3	.82	50	30	1.00
5	4	.92	120	25	1.00
5	10	1.00	140	25	.98
25	3	.46	160	25	.93
25	5	.68	180	25	.86
25	10	.95	200	25	.77
25	30	1.00	220	25	.71
50	3	.32	240	25	.68
50	5	.52	260	25	.66
50	10	.84	280	25	.64

As one would expect, $\sigma(n, T, 0)$ is an increasing function in n, for fixed T, and a decreasing function in T, for fixed n. In some cases, the values of $\sigma(n, T, 0)$ serve as convenient lower bounds for the distribution of total claims. For example,

$$F(4, 5) = P\left[\sum_{i=1}^{N(5)} X_i < 4\right]$$

$$\geq P\left[\max_{0 \leq t \leq 5} \sum_{i=1}^{N(t)} X_i < 4\right] = .92.$$

Similarly, $\quad F(10, 25) = P\left[\sum_{i=1}^{N(25)} X_i < 10\right] \geq .95.$

COROLLARY 1. If $P[X \leq 0] = 0$, and $E[X] < \infty$,

$$\int_0^\infty \int_0^\infty e^{-wT - Zu} d_u \sigma(u, T, p_1 + \lambda) dT = \frac{1 - Z/y(w)}{w - \int_0^\infty e^{-Zx} dP(x) + 1 - Z(p_1 + \lambda)}$$

---

[1] Reprinted with the permission of the editors of *Skandinavisk Aktuarietidskrift*.

where $y(w)$ is the only non-negative solution of

$$w = (p_1 + \lambda) y(w) + \int_0^\infty e^{-y(w)x} dP(x) - 1, \quad w \geqslant 0.$$

Exercise. Solve this integral equation for $y(w)$ when $P(x) = 1 - e^{-x}, x \geqslant 0$.

COROLLARY 2. *If $P[X \leqslant 0] = 0$, and $E[X] < \infty$,*

$$F(x, T) = \lim_{\delta \to 0+} \sigma(x - \delta, T, \delta/T)$$

where the σ functions are determined by Corollary 1.

THEOREM 2. (See Cramér [21], p. 20.) *If $p_1 < \infty$,*

$$F(x, t) = \sum_{n=0}^\infty \frac{t^n e^{-t}}{n!} P^{n*}(x)$$

where
$$P^{0*}(x) = \begin{cases} 0, & x < 0 \\ 1, & x \geqslant 0 \end{cases}$$

$$P^{1*}(x) = P(x)$$

$$P^{n*}(x) = \int_{-\infty}^\infty P^{(n-1)*}(x - z) \, dP(z), \quad n > 1.$$

Proof.
$$F(x, t) = P\left[\sum_{i=1}^{N(t)} X_i \leqslant x\right]$$

$$= \sum_{n=0}^\infty P\left[\sum_{i=1}^{N(t)} X_i \leqslant x \,\middle|\, N(t) = n\right] P[N(t) = n]$$

$$= \sum_{n=0}^\infty P\left[\sum_{i=1}^n X_i \leqslant x\right] P[N(t) = n]$$

$$= \sum_{n=0}^\infty P^{n*}(x) \frac{e^{-t} t^n}{n!}.$$

Remark. In the above, if $N(t) = 0$, we interpret the sum to equal 0.

$$P\left[\sum_{i=1}^n X_i \leqslant x\right] = P^{n*}(x),$$

the n-fold convolution of $P(x)$ with itself, in view of the fact that the X_i's are independent, identically distributed random variables.

Let us now consider some examples of Theorem 2. For ease of notation, let

$$x(t) = \sum_{i=1}^{N(t)} X_i.$$

Let
$$F(z) = \begin{cases} 0, & z < 1 \\ 1, & z \geqslant 1 \end{cases}.$$

Here one might represent $ 5,000.00.

Then
$$P^{2*}(x) = \int_0^\infty P^{1*}(x-z)\,dF(z)$$
$$= P^{1*}(x-1)$$
$$= \begin{cases} 0, & x < 2 \\ 1, & x \geqslant 2 \end{cases}$$

and for $n = 3, 4, \ldots$

$$P^{n*}(x) = \begin{cases} 0, & x < n \\ 1, & x \geqslant n \end{cases}.$$

Thus
$$F(x, t) = P[x(t) \leqslant x]$$
$$= \sum_{n=0}^{\infty} \frac{e^{-t} t^n}{n!} P^{n*}(x)$$
$$= \sum_{n=0}^{[x]} \frac{e^{-t} t^n}{n!}.$$

For this distribution of claims, $x(t)$ can only assume integral values. Thus for k a positive integer $\geqslant 1$

$$P[x(t) = k] = P[x(t) \leqslant k] - P[x(t) \leqslant k-1]$$
$$= \frac{e^{-t} t^k}{k!}.$$

Of course,
$$P[x(t) = 0] = P[N(t) = 0]$$
$$= e^{-t}.$$

We can thus assign values to the joint probabilities. Thus assume that $0 < t_1 < t_2 < t_3$ and x_1, x_2, x_3 are positive integers with $x_1 < x_2 < x_3$. Then

$P[x(t_1) = x_1,\ x(t_2) = x_2,\ x(t_3) = x_3]$
$= P[x(t_1) = x_1,\ x(t_2) - x(t_1) = x_2 - x_1,\ x(t_3) - x(t_2) = x_3 - x_2]$
$= P[x(t_1) = x_1] P[x(t_2) - x(t_1) = x_2 - x_1] P[x(t_3) - x(t_2) = x_3 - x_2]$

by independent increments as we shall see

$$= P[x(t_1) = x_1]P[x(t_2-t_1) = x_2-x_1]P[x(t_3-t_2) = x_3-x_2]$$

by stationary increments as we shall see

$$= \frac{e^{-t_1}t_1^{x_1}}{x_1!} \frac{e^{-(t_2-t_1)}(t_2-t_1)^{(x_2-x_1)}}{(x_2-x_1)!} \frac{e^{-(t_3-t_2)}(t_3-t_2)^{(x_3-x_2)}}{(x_3-x_2)!}.$$

Let
$$F(x) = \begin{cases} 0, & x<0 \\ 1-e^{-x}, & x \geq 0 \end{cases}.$$

$$P^{2*}(x) = \int_0^\infty P^{1*}(x-z)\,dF(z)$$

$$= \int_0^x [1-e^{-(x-z)}]e^{-z}dz$$

$$= 1 - \sum_{n=0}^1 \frac{e^{-x}x^n}{n!}$$

$$= \int_0^x \frac{e^{-z}z}{1!}dz.$$

By math. induction,

$$P^{n*}(x) = \int_0^x \frac{e^{-z}z^{n-1}}{(n-1)!}dz,$$

the incomplete gamma function. The calculation of $P^{n*}(x)$ appears on p. 417 of [28].

Thus
$$F(x,t) = P[x(t) \leq x]$$

$$= \sum_{n=0}^\infty \frac{e^{-t}t^n}{n!} P^{n*}(x)$$

and the density function would be (for $x>0$)

$$f(x,t) = \frac{dF(x,t)}{dx} = \sum_{n=1}^\infty \frac{e^{-t}t^n}{n!} \frac{e^{-x}x^{n-1}}{(n-1)!}.$$

Note that
$$\int_0^\infty f(x,t)\,dx = 1-e^{-t}.$$

Thus,
$$F(x,t) = e^{-t} + \int_0^x f(w,t)\,dw$$

for $x \geqslant 0$.

Now assume $t_2 > t_1$ and consider the increment $x(t_2) - x(t_1)$.

$$P[x(t_2) - x(t_1) \leqslant y] = P\left[\sum_{i=0}^{N(t_2)} X_i - \sum_{i=0}^{N(t_1)} X_i \leqslant y\right]$$

$$= P\left[\sum_{i=N(t_1)+1}^{N(t_2)} X_i \leqslant y\right]$$

$$= P\left[\sum_{i=1}^{N(t_2-t_1)} X_i \leqslant y\right] = F(y, t_2 - t_1)$$

by our assumptions and the fact that

$$P[N(t) - N(s) = k] = P[N(t-s) = k].$$

For more general distribution functions, the formula in Theorem 2 is of little value. Therefore, it is very important to study a very effective approximation to $F(x,t)$ developed by Newton Bowers (see [17]). It uses the moments of the random variable $\sum_{i=1}^{N(t)} X_i$ and the incomplete gamma function. We will describe the approximation more fully in the next section.

COROLLARY 3. *If* $P[X \leqslant 0] = 0$, *and* $E[X] < \infty$,

$$\int_0^\infty e^{-u\alpha} d_u \sigma(u, \infty, p_1 + \lambda) = \frac{\alpha\lambda}{\int_0^\infty e^{-\alpha x} dP(x) - 1 + \alpha(p_1 + \lambda)}$$

COROLLARY 4. *If* $u \geqslant 0$,

$$\psi(u) = 1 - \lambda I_\alpha\left[\frac{1}{\int_0^\infty e^{-\alpha x} dP(x) - 1 + \alpha(p_1 + \lambda)}\right]$$

where $I_\alpha\{f(\alpha)\}$ denotes the inverse Laplace transform of $f(\alpha)$.

Proof. In Corollary 3 let $\psi^*(u) = \sigma(u, \infty, p_1 + \lambda)$ and integrate by parts to obtain

$$\int_0^\infty e^{-\alpha u} d_u \psi^*(u) = -\psi^*(0) + \alpha \int_0^\infty e^{-\alpha u} \psi^*(u)\,du.$$

Since
$$\psi^*(u) = \lim_{T \to \infty} P\left[\left\{\text{maximum}_{0 \leq t \leq T} \sum_{i=1}^{N(t)} X_i - t(p_1 + \lambda)\right\} < u\right]$$

and the quantity in brackets equals 0 for $t=0$, the maximum is ≥ 0 and hence $\psi^*(0) = 0$. Dividing by α, we obtain

$$\int_0^\infty e^{-\alpha u} \psi^*(u) \, du = \frac{\lambda}{\int_0^\infty e^{-\alpha x} dP(x) - 1 + \alpha(p_1 + \lambda)}.$$

Inversion gives $\psi^*(u)$, and then for $u > 0$, $\psi(u) = 1 - \psi^*(u)$. Since $\psi(u)$ is continuous for $u > 0$ and continuous from the right at $u = 0$, the result holds for $u = 0$ also, and gives the usual $p_1/(p_1 + \lambda)$ value.

The problems are now reduced to inverting some transforms. There has been some success in this regard.

Example 1. (See [8].) $P[X \leq z] = 1 - e^{-Az}$, $z \geq 0$, $A > 0$. Then

$$\sigma\left(u, \infty, \frac{1}{A} + \lambda\right) = 1 - \frac{1}{1 + \lambda A} \exp\left[\frac{-\lambda A^2}{1 + \lambda A} u\right].$$

If we use $1,000 units and let $A = 1$, and let $\psi(u) = 1 - \sigma(u, \infty, 1 + \lambda)$, we obtain the following table: ([1])

$\psi(u) = .01$		$\psi(u) = .05$		$\psi(u) = .1$	
u	λ	u	λ	u	λ
18.800	.3	11.833	.3	8.831	.3
26.483	.2	16.846	.2	12.695	.2

For $A = .1$, the u's are all 10 times bigger.

Exercise 1. Verify example 1, using the inverse Laplace transforms in section 2.7. Hint: Use partial fractions on the expression to be inverted.

Exercise 2. What conditions are needed on A, B, C, D, E, and F in order for

$$Q(x) = \begin{cases} 0, & x < 0 \\ 1 - Ae^{-Bx} - Ce^{-Dx} - Ee^{-Fx}, & x \geq 0 \end{cases}$$

to be a distribution function?

([1]) This table is reproduced with the permission of the editors of *Transactions, Society of Actuaries*.

Exercise 3. Assume that we need to know $\psi(u)$ for the following step function.

$$P(x) = \begin{cases} 0, & x < 1 \\ .4, & 1 \leq x < 5 \\ .6, & 5 \leq x < 10 \\ .8, & 10 \leq x < 25 \\ .9, & 25 \leq x < 50 \\ 1.0, & 50 \leq x. \end{cases}$$

It is frankly very difficult to do so. Since $\psi(u)$ is only used to guide the actuary's judgment, assume that we will accept a conservative approximation to $\psi(u)$.

Choose $A, B, C, D, E,$ and F so that $Q(1) \leq P(1)$, $Q(5) \leq P(5)$, $Q(10) \leq P(10)$, $Q(25) \leq P(25)$, $Q(50) \leq P(50)$ but so that the values are as close together as possible. This will force $Q(x) \leq P(x)$, $0 \leq x < \infty$, and $Q(x)$ is a slightly more dangerous distribution than $P(x)$ because

$$P[\text{Claim} > x] = 1 - P(x) \leq 1 - Q(x).$$

If possible, let $\qquad C$ or E equal 0.

Since $Q(x)$ is a slightly more dangerous distribution than $P(x)$, if

$$\psi_Q(u) \leq k \text{ for } Q(x), \text{ then } \psi_P(u) \leq k \text{ for } P(x).$$

Exercise 4. Using Corollary 4, derive an expression for $\psi(u)$ for the $Q(x)$ chosen in #3.

Exercise 5. What value of u will make $\psi(u) \doteq .01$ if $\lambda = .3p_1$? Remember that for that u, $\psi(u, T) \leq .01$ for any $T > 0$ since we always have $\psi(u, T) < \psi(u)$.

Example 2.

$$P[X \leq z] = \begin{cases} 0, & z < 1 \\ 1, & z \geq 1 \end{cases}$$

$$1 - \psi(u) = \sigma(u, \infty, 1+\lambda) = \frac{\lambda}{1+\lambda} \sum_{n=0}^{[u]} \frac{(-1)^n (u-n)^n}{n!(1+\lambda)^n} \exp\left[\frac{u-n}{1+\lambda}\right]$$

A reference for this example and a table is given in [7].

Probability of Ruin, $\psi(u)$[1]

u/λ	.01	.02	.03	.04	.05	.06
1	.973	.948	.923	.899	.877	.855
2	.955	.912	.872	.834	.798	.764
3	.936	.877	.822	.771	.724	.681
4	.918	.843	.775	.713	.657	.606
5	.899	.810	.731	.660	.597	.540
6	.882	.779	.689	.610	.542	.481
7	.864	.749	.650	.565	.492	.429
8	.847	.720	.612	.522	.446	.382
9	.831	.692	.577	.483	.405	.341
10	.814	.665	.544	.447	.368	.303
11	.798	.639	.513	.413	.334	.270
12	.783	.615	.484	.382	.303	.241
13	.767	.591	.456	.354	.275	.215
14	.752	.568	.430	.327	.250	.191
15	.737	.546	.406	.303	.227	.170
16	.723	.525	.383	.280	.206	.152
17	.709	.504	.361	.259	.187	.135
18	.695	.485	.340	.240	.170	.120
19	.681	.466	.321	.222	.154	.107
20	.668	.448	.302	.205	.140	.096

[1] This table is reproduced with the permission of the editors of *Skandinavisk Aktuarietidskrift*.

Example 3. Same distribution as Example 1. (See [8]) for a derivation of this through Corollary 2.

$$F(x, T) = P\left[\sum_{i=1}^{N(T)} X_i \leq x\right] = e^{-T}\left[1 + \int_0^x e^{-wA} TA \sum_{y=0}^{\infty} \frac{(TAw)^y}{(y+1)!\, y!}\, dw\right]$$

This example will be resumed in Section 3.6.

Luckily for research, actuaries were not happy with these examples. Their problems involved more complicated distributions. The Laplace transforms could not be inverted. However, the Bowers paper about the approximation of distributions appeared helpful. It motivated the following theorems which appear in [9]. See also [18] and [10].

Let
$$Z = \sup_{0 \leq t < \infty}\left[\sum_{i=1}^{N(t)} X_i - t(p_1 + \lambda)\right].$$

We have used sup rather than max because it could equal $+\infty$. If

$$Z_T = \max_{0 \leqslant t \leqslant T} \left(\sum_{i=1}^{N(t)} X - t(p_1 + \lambda) \right),$$

then $\lim_{T \to \infty} Z_T(w) = Z(w)$ for every $w \in \Omega$, the sample space described by Cramér. Cramér shows that Z is a measurable function with respect to a σ-field of the subsets of the sample space. (See [21], pages 27, 50–52.)

THEOREM 3. *If the $\{X_i\}$ are such that $P[X_i \leqslant 0] = 0$ and $E(X^2) < \infty$, then $E(Z) = E(X^2)/(2\lambda)$.*

Proof. Since $Z \geqslant 0$,

$$E(Z) = \int_0^\infty u \, d_u P_z(u) = \int_0^\infty u \, d_u \psi^*(u).$$

By Corollary 4,

$$\int_0^\infty e^{-\alpha u} d_u \psi^*(u) = \frac{\alpha \lambda}{\int_0^\infty e^{-\alpha x} dP(x) - 1 + \alpha(p_1 + \lambda)}.$$

One now uses the Stieltjes version of Theorem 14 of Widder [43], page 358, to obtain

$$E(Z) = -\frac{d}{d\alpha} \int_0^\infty e^{-\alpha u} d_u \psi^*(u) \Big|_{\alpha=0}$$

$$= \int_0^\infty e^{-\alpha u} u \, d_u \psi^*(u) \Big|_{\alpha=0}.$$

Differentiating the right-hand side of the expression for

$$\int_0^\infty e^{-\alpha u} d_u \psi^*(u)$$

produces the form $0/0$ when $\alpha = 0$. One then applies L'Hospital's rule twice to obtain the result.

Exercise 6. Carry out the differentiation and the application of L'Hospital's rule.

THEOREM 4. If the $\{X_i\}$ are such that $P[X_i \leqslant 0]=0$ and $E(X^3)<\infty$,

$$\operatorname{Var}(Z) = \frac{E(X^3)}{3\lambda} + [E(Z)]^2.$$

Proof. $\operatorname{Var}(Z) = E(Z^2) - [E(Z)]^2$.

$$E(Z^2) = \int_0^\infty u^2 d_u \psi^*(u)$$

$$= \frac{d^2}{d\alpha^2} \int_0^\infty e^{-\alpha u} d_u \psi^*(u) \big|_{\alpha=0}.$$

Differentiating the right-hand side of the expression for

$$\int_0^\infty e^{-\alpha u} d_u \psi^*(u)$$

twice produces the form $0/0$ when $\alpha = 0$. One then applies L'Hospital's rule three times and does much elementary algebra to obtain the result.

THEOREM 5. If the $\{X_i\}$ are such that $P[X_i \leqslant 0]=0$ and $E(X^3)<\infty$,

$$P[Z \leqslant u] \doteq \frac{\lambda}{p_1+\lambda} + \frac{p_1}{p_1+\lambda} \int_0^u \frac{t^{\alpha-1} e^{-t/\beta}}{\Gamma(\alpha) \beta^\alpha} dt$$

where

$$\beta = \tfrac{2}{3} \frac{E(X^3)}{E(X^2)} + \frac{E(X^2)}{2\lambda}\left(1 - \frac{\lambda}{p_1}\right) \quad \text{and} \quad \alpha = \frac{E(X^2)}{2\lambda} \frac{p_1+\lambda}{p_1} \bigg/ \beta.$$

Remark: $\quad \psi(u) = 1 - P[Z \leqslant u] = \frac{p_1}{p_1+\lambda} \int_u^\infty \frac{t^{\alpha-1} e^{-t/\beta}}{\Gamma(\alpha) \beta^\alpha} dt.$

Proof. One approximates the distribution of Z by a mixed type distribution which consists of a lump of probability at the origin and a complementary multiple of an incomplete gamma distribution whose first two moments are found from $E(Z)$ and $E(Z^2)$.

Thus $$P[Z=0] = \frac{\lambda}{p_1+\lambda}$$

and we approximate $P[Z>u]$ as

$$P[Z>u] \doteq \frac{p_1}{p_1+\lambda} \int_u^\infty \frac{t^{\alpha-1} e^{-t/\beta}}{\Gamma(\alpha) \beta^\alpha} dt \quad \text{for} \quad u \geqslant 0.$$

To determine α and β we have the equations

$$E(Z) = \frac{\lambda}{p_1 + \lambda}(0) + \frac{p_1}{p_1 + \lambda}\int_{0+}^{\infty} t \frac{t^{\alpha-1} e^{-t/\beta}}{\Gamma(\alpha)\beta^\alpha} dt = \frac{p_1}{p_1 + \lambda}\alpha\beta,$$

and

$$E(Z^2) = \frac{p_1}{p_1 + \lambda}(\alpha(\alpha+1)\beta^2).$$

From Theorems 3 and 4, and the method of moments, one obtains

$$\frac{p_2}{2\lambda} = \frac{p_1}{p_1 + \lambda}\alpha\beta$$

and

$$\frac{p_3}{3\lambda} + 2\left(\frac{p_2}{2\lambda}\right)^2 = \frac{p_1}{p_1 + \lambda}(\alpha(\alpha+1)\beta^2).$$

Substituting

$$\alpha\beta = \frac{p_2}{2\lambda}\frac{p_1 + \lambda}{p_1}$$

in the second equation gives β and then α is obtained.

We will now discuss the accuracy of the method.

Example 1. $P(X \leq z) = 1 - e^{-z}$, $z \geq 0$. Note that

$$E(X^k) = k!, \quad E(Z) = \frac{1}{\lambda},$$

$$\text{Var}(Z) = \frac{2}{\lambda} + \frac{1}{\lambda^2}, \quad E(Z^2) = \frac{2(1+\lambda)}{\lambda^2}, \quad \alpha = 1,$$

and

$$\beta = \frac{1+\lambda}{\lambda}.$$

Therefore,

$$\psi(u) = P[Z > u] \div \frac{1}{1+\lambda}\int_u^\infty \frac{e^{-t/\beta}}{\beta} dt = \frac{1}{1+\lambda} e^{-u/\beta} = \frac{1}{1+\lambda} e^{-u\lambda/(1+\lambda)} \quad \text{for} \quad u \geq 0.$$

Since this is also the exact answer, the approximation is 100 % accurate here.

Exercise 7. Verify that $E(Z)$, Var (Z), α, and β are as stated.

Example 2. (See Cramér [21], p. 43). Assume the density is given by

$$P'(z) = \begin{cases} A e^{-\alpha z} + B(z+b)^{-\beta}, & 0 < z < 500 \\ 0, & z > 500 \end{cases}$$

where $A = 4.897954$, $\alpha = 5.514588$, $B = 4.503$, $b = 6$, and $\beta = 2.75$. Here $p_1 = 1$. This density was based on an experience from Swedish non-industry fire insurance for the years 1948–1951.

A table giving approximate values for $\psi(u)$ for this example will be found after the following material on the exact solution, and the Lundberg approximation method.

We will need the following theorem for the exact results. (This theorem is in Cramér [21], pp. 52 and 61.)

THEOREM 6. Assume that the $\{X_i\}$ have a common distribution $P(y)$ which satisfies

$$\int_{-\infty}^{0} |y| \, dP(y) < \infty,$$

and

$$\int_{0}^{\infty} e^{\sigma y} \, dP(y) < \infty$$

for some $\sigma > 0$. (Except for pathological cases skewed to the left of 0, this requires all moments to be finite.) Let $\psi(u) = 1 - \sigma(u, \infty, p_1 + \lambda)$. Then $\psi(u)$ satisfies the integral equation

$$(p_1 + \lambda) \psi(u) = \int_{u}^{\infty} Q(v) \, dv + \int_{0}^{\infty} \psi(v) Q(u-v) \, dv$$

where

$$Q(u) = \begin{cases} -P[X \leq u], & u < 0 \\ 1 - P[X \leq u], & u \geq 0. \end{cases}$$

Note that $Q(u) = 0$ for $u < 0$ in our example. The table which appears later includes exact values of $\psi(u)$ obtaining by solving this integral equation with the aid of an electronic computer.

We will also need the Lundberg asymptotic approximation to $\psi(u)$, as $u \to \infty$. This appears on page 45 of [21].

THEOREM 7. Assume that the $\{X_i\}$ have a common distribution $P(y)$ which satisfies $\int_{0}^{\infty} e^{\sigma y} \, dP(y) < \infty$ for some $\sigma = R + \theta$ with $\theta > 0$ where R

is the only positive root of the equation

$$1 + (p_1 + \lambda)s - \int_0^\infty e^{sy} dP(y) = 0.$$

Also assume that $P(0) = 0$, that is, that all risk sums are positive. Then, as $u \to \infty$,

$$\psi(u) = \frac{\lambda}{p'(R) - p_1 - \lambda} e^{-Ru} + O(e^{-(R+\theta)u})$$

where

$$p(\theta) = \int_0^\infty e^{\theta y} dP(y),$$

and

$$p'(R) = \frac{d}{d\theta} p(\theta)\big|_{\theta=R}.$$

In almost all practical situations

$$p'(R) = \int_0^\infty y e^{Ry} dP(y).$$

Before resuming our discussion of the incomplete gamma function approximation to $\psi(u)$, let us consider some examples of the Lundberg method. The first two examples come from pages 186–187 of [8], as corrected on page 279 of [10].

Example 1.

$$P_1(y) = \begin{cases} .0, & y < 2, \\ .3, & 2 \leqslant y < 5, \\ .5, & 5 \leqslant y < 10, \\ .8, & 10 \leqslant y < 20, \\ 1.0, & y \geqslant 20. \end{cases}$$

Then $p_1 = 8.6$ ($\$8\,600$, since we are using $\$1,000$ units. If we let $\lambda = .3 p_1 \doteq 2.6$, then the equation $1 + 11.2s - .3e^{2s} - .2e^{5s} - .3e^{10s} - .2e^{20s} = 0$ has an approximate root of $R \doteq .035$. Hence

$$C = \frac{\lambda}{p'(R) - p_1 - \lambda} = 0.8823$$

and $\psi(\$128\,000) \doteq .01$.

Example 2.

$$P_2(y) = \begin{cases} 0, & y < 2, \\ .3, & 2 \leq y < 5, \\ .5, & 5 \leq y < 10, \\ .8, & 10 \leq y < 20, \\ .85, & 20 \leq y < 30, \\ .90, & 30 \leq y < 40, \\ .95, & 40 \leq y < 50, \\ 1.00, & y \geq 50. \end{cases}$$

As expected, p_1 is now larger, and has the value 11.6. If we again let $\lambda = .3 p_1$, which is now approximately 3.5, then $R \doteq .0175$. Now $C = 0.8424$ and in order to keep the same small probability of ruin, it takes considerably more initial capital, i.e. $\psi(\$253,143) \doteq .01$.

Example 3.

$$P(x) = \begin{cases} 1 - e^{-x}, & x \geq 0 \\ 0, & x < 0. \end{cases}$$

Then R is the only positive root of the equation

$$1 + (1+\lambda)R - \frac{1}{1-R} = 0.$$

Hence $R = \lambda/(1+\lambda)$. Furthermore, $p(\theta) = (1-\theta)^{-1}$, $p'(R) = (1-R)^{-2}$, and

$$\frac{\lambda}{p'(R) - p_1 - \lambda} = \frac{1}{1+\lambda}.$$

Thus, for u large,

$$\psi(u) \doteq \frac{1}{1+\lambda} \exp\left\{-\frac{\lambda u}{1+\lambda}\right\}.$$

This is actually the exact answer.

A summary of the exact values of $\psi(u)$, Lundberg's asymptotic values for $\psi(u)$, and the Beekman–Bowers values for $\psi(u)$ for the Swedish non-industry fire insurance density will now be given. This is given on page 279 of [10], and is reprinted below.

Fire insurance distribution: values of the ruin [1] probability $\psi(u)$ for $\lambda = 0.3$

u	Exact	Lundberg's approximation	Ratio Lundberg/Exact	Beekman–Bowers Approximation	Ratio B-B./Exact
20	0.5039	0.4524	0.898	0.5141	1.020
40	.3985	.3904	.980	.4098	1.028
60	.3280	.3370	1.027	.3367	1.027
80	.2757	.2909	1.055	.2811	1.020
100	0.2346	0.2511	1.070	0.2370	1.010

[1] Reprinted with the permission of the editors of *Transactions, Society of Actuaries*.

Exercise 8. Verify that p_1, R, and C are as stated in Examples 1, 2, and 3 of the Lundberg approximation.

Exercise 9. Refer back to exercises 3, 4, and 5 following Corollary 4. Approximate $\psi(u)$ for $P(x)$ through the gamma method. What value makes $\psi(u) \doteq .01$? Compare this with the $Q(x)$ calculation. Obviously $\psi_{P(x)}(u) \leqslant \psi_{Q(x)}(u)$.

Several hints should be given on the use of tables of the incomplete gamma function. Using the transformation $w = t/\beta$,

$$\int_u^\infty \frac{t^{\alpha-1} e^{-t/\beta}}{\beta^\alpha} dt = \int_{u/\beta}^\infty w^{\alpha-1} e^{-w} dw.$$

Furthermore,

$$\Gamma(\alpha) = \int_0^\infty w^{\alpha-1} e^{-w} dw = \int_0^{u/\beta} w^{\alpha-1} e^{-w} dw + \int_{u/\beta}^\infty w^{\alpha-1} e^{-w} dw.$$

Hence we can express

$$\int_u^\infty \frac{t^{\alpha-1} e^{-t/\beta}}{\Gamma(\alpha) \beta^\alpha} dt = 1 - \int_0^{u/\beta} \frac{w^{\alpha-1} e^{-w} dw}{\Gamma(\alpha)}.$$

Tables of the incomplete gamma function will be found in Pearson [30], and Salvosa [32]. Three more references are now available and are very useful. The *Handbook of Mathematical Functions*, edited by M. Abramowitz and I. Segun, Dover Publications, New York, 1968, [1], discusses the incomplete gamma function on page 260 (section 6.5), pages 940–941 (sections 26.4.1 and 26.4.20) and tables it on pages 978–983. Another reference is "New Tables of the Incomplete Gamma-Function Ratio and

of Percentage Points of the Chi-Square and Beta Distributions" by H. Leon
Harter, U.S. Aerospace Research Laboratories, U.S. Government Printing
Office, Washington, D. C., 1964, [26]. A brief and effective program for
a mini-computer will be found in an article by Gordon D. Shellard which
appears in Issue 1972.1 of ARCH (Actuarial Research Clearing House),
[37]. In their paper, Grandell and Segerdahl compare the Cramér–
Lundberg, the Beekman–Bowers, and the Bohman (see [16]) approximations to $\psi(u)$ in the case that the claim distribution is the Γ-distribution
with parameter b. For $b=1$, the Γ-distribution is the negative exponential
distribution. Except for $b=1$, the Cramér–Lundberg and Bohman approximations give values which are closer to the exact values than those
furnished by the Beekman–Bowers method. An exact formula derived
by Olof Thorin is given. In the notation of [30] (see equation (xii) on
page vii),

$$\int_u^\infty \frac{t^{\alpha-1} e^{-t/\beta}}{\Gamma(\alpha)\beta^\alpha} dt = 1 - I\left(\frac{u}{\beta\sqrt{\alpha}}, \alpha-1\right).$$

A few words should be said about interpolating for values of $I(u, p)$. Let
us assume that one needs $I(4.73, 22.3)$, for example. It is easy to find
$I(4.7, 22.2)$, $I(4.7, 22.4)$, $I(4.8, 22.2)$, and $I(4.8, 22.4)$. For simplicity,
label them a, b, c, d, respectively. First form $I(4.7, 22.3) = .5a + .5b = e$
and $I(4.8, 22.3) = .5c + .5d = f$. Finally, $I(4.73, 22.3) = .7e + .3f$. This suggested method is a two-way linear interpolation method.

We will now give three applications of the use of $\psi(u)$.

Application 1

In this application, the company is considering raising its retention level
from $20 000.00 to $50 000.00. The following show the current and
projected distribution of claims. One unit is $1 000.00. The current
distribution has weights of .30, .20, .30, and .20 at $2 000, $5 000, $10 000,
and $20 000. It is projected that a $50 000 retention level would spread the
$20 000 policies evenly among $20 000, $30 000, $40 000, and $50 000
face amounts. For future reference, we label these distributions:

Example 3. $P[X \leq z] = $ 0 for $z < 2$ $P[X] \leq z] = 0,\quad z < 2$
.3, $2 \leq z < 5$ .3, $2 \leq z < 5$
.5, $5 \leq z < 10$ .5, $5 \leq z < 10$
.8, $10 \leq z < 20$ .8, $10 \leq z < 20$
1.0, $20 \leq z$ .85, $20 \leq z < 30$
                                                  .90, $30 \leq z < 40$
                                                  .95, $40 \leq z < 50$
We again let $\lambda = .3 p_1$ in each case.   1.00, $50 \leq z$

What amount u of initial capital is needed to hold $\psi(u) \leq .01$?

For distribution 1, $u = \$125\,000.00$.

For distribution 2, $u = \$250\,000.00$.

Thus, increasing the retention limit from $\$20\,000.00$ to $\$50\,000.00$ requires an additional $\$125\,000.00$ of initial capital. You can also solve for the retention limit knowing the upper limit on adverse fluctuation.

Application 2

Assume the company is contemplating entering a new line, say accident and sickness insurance. Your company would like to set aside some capital for this venture. How much is needed in order that there is a 99% chance that you will not have to dip into your other funds? Solution: Use $\psi(u)$.

Application 3

An interesting theoretical problem for any company would be to divide the surplus according to the needs of each separate line of business. If one uses $\psi(u)$ for each line, the sum of the surplus requirements exceeds that obtained for the entire company. This reflects the fact that unfavourable experience in one line may be balanced by favorable experience in another line. William Frye has published a lengthy paper on this problem, with extensive tables. See [23].

Approximation of $\psi(u, T)$. Let

$$\psi(u, T) = P\left[\max_{0 \leq t \leq T} \left(\sum_{i=1}^{N(t)} X_i - (p_1 + \lambda) t\right) > u\right],$$

the probability of being ruined before time T. Analytic expressions for $\psi(u, T)$ are almost non-existent. The excellent paper by C. O. Segerdahl [35] does contain one example for which $\psi(u, T)$ is known. Recently, Hilary Seal presented a paper in which an approximation to $\psi(u, T)$ is achieved using electronic computer simulation techniques [33]. This will be described in a later section. Olof Thorin reported on the numerical inversion of the double transform for $\psi(u, T)$ in [40], and this method was exemplified by Nils Wikstad in [44]. See also [41]. This method will be considered in section 3.5. Newton Bowers and the author have derived an approximation for $\psi(u, T)$ in [11] and [12]. This will be briefly described in section 3.4.

Other References

Mention should also be made of Hilary Seal's book *Stochastic Theory of a Risk Business*, Wiley, New York, 1969 [34], which gives a thorough treatment of this subject, the book *Risk Theory* by Beard, Pentikainen, and Pesonen [6], and the paper "A Review of the Collective Theory of Risk" by Carl Philipson, *Skandinavisk Aktuarietidskrift*, 1968, pages 1–41, which contains 364 references! [31]. A recent book in the area of risk theory is *Mathematical Methods in Risk Theory* by Hans Bühlmann, [20].

The reader's attention is drawn to the *Proceedings, Wisconsin Actuarial Conference*, to appear in 1974. The paper [24] by Hans Gerber discusses the implications of allowing an interest assumption to be used in collective risk theory. The paper [27] by C. J. Jackson examines various models which treat the insurance company as operating under the joint influence of claim fluctuation and investment fluctuation.

3.1. Approximations to $F(x, t)$

Before discussing these approximations, we will need some preliminary results. We will assume that the X random variables representing the claims have a common moment generating function

$$M(\theta) = \int_{-\infty}^{\infty} e^{\theta x} dP(x).$$

Then $M(0) = 1$, $M^1(0) = p_1$, and $M''(0) = p_2$.

We can now calculate the moment generating function for the random variable

$$x(t) = \sum_{i=1}^{N(t)} X_i, \quad 0 \leqslant t < \infty.$$

As a preliminary remark, let us recall that $P^{n*}(x)$ is the distribution of the nth partial sum $\sum_{i=1}^{n} X_i$, and by the independence of the X_i's,

$$E\left[\exp\left(\theta \sum_{i=1}^{n} X_i\right)\right] = \int_{-\infty}^{\infty} e^{\theta x} dP^{n*}(x) = [M(\theta)]^n.$$

Hence
$$E\{e^{\theta x(t)}\} = \int_{-\infty}^{\infty} e^{\theta x} d_x F(x, t)$$

$$= \sum_{n=0}^{\infty} \frac{t^n e^{-t}}{n!} \int_{-\infty}^{\infty} e^{\theta x} dP^{n*}(x)$$

$$= \sum_{n=0}^{\infty} \frac{t^n e^{-t}}{n!} [M(\theta)]^n$$

$$= \exp\{t[M(\theta) - 1]\}.$$

The following is an alternative derivation of the moment generating function for $x(t)$.

$$E\left[e^{\theta} \sum_{i=1}^{N(t)} X_i\right] = \sum_{n=0}^{\infty} E\left[e^{\theta} \sum_{i=1}^{N(t)} X_i \,\Big|\, N(t) = n\right] P[N(t) = n]$$

$$= \sum_{n=0}^{\infty} E\left[e^{\theta} \sum_{i=1}^{n} X_i\right] P[N(t) = n]$$

$$= \sum_{n=0}^{\infty} [M(\theta)]^n \frac{e^{-t} t^n}{n!}$$

$$= e^{-t} \sum_{n=0}^{\infty} \frac{[tM(\theta)]^n}{n!}$$

$$= e^{t[M(\theta)-1]}.$$

Hence
$$E\{x^i(t)\} = \frac{d^i}{d\theta^i} \exp\{t[M(\theta)-1]\}\big/_{\theta=0}$$

and it is easy to verify that

$$E\{x(t)\} = p_1 t \quad \text{and} \quad E\{x^2(t)\} = p_2 t + p_1^2 t^2.$$

Thus
$$\text{Var}\{x(t)\} = p_2 t \quad \text{and Std. Dev.} \quad \{x(t)\} = \sqrt{p_2 t}.$$

As early as 1903, F. Lundberg showed that

$$\lim_{t \to \infty} P\{x(t) > p_1 t - \beta\sqrt{p_2 t}\} = \int_{-\beta}^{\infty} \frac{1}{\sqrt{2\pi}} e^{-y^2/2} dy = \frac{1}{\sqrt{2\pi}} \int_{-\infty}^{\beta} e^{-y^2/2} dy$$

for any fixed value of x. However, Lundberg observed that the approximation of replacing the probability by the normal integral usually was not accurate enough for practical purposes.

Since
$$P\{x(t) > p_1 t - \beta\sqrt{p_2 t}\}$$
$$= P\left\{\frac{x(t) - p_1(t)}{\sqrt{p_2 t}} > -\beta\right\},$$

if we imagine $x(t)$ to be $\sum_{i=1}^{t} X_i$, rather than $\sum_{i=1}^{N(t)} X_i$, and think of t as assuming positive integral values, the above asymptotic result would follow directly from the Central Limit Theorem. However, it is rash to think of $N(t)$ as t because even for $t=1\,000$, there is an approximate .997 probability that $N(t)$ will be between $1\,000-95$ and $1\,000+95$. That is,

$$P[1\,000 - 3\sqrt{1\,000} < N(t) < 1\,000 + 3\sqrt{1\,000}]$$

$$\doteq \frac{1}{\sqrt{2\pi}} \int_{-3}^{3} e^{-x^2/2} dx = .997, \quad \text{and} \quad 3\sqrt{1\,000} \doteq 95.$$

H. Cramér developed certain asymptotic expansions derived from the normal approximation, and derived methods for computing the errors in the various approximations. References 18, 19, and 20 of [21] discuss these matters in detail. As explained on pages 31–33 of [21],

$$P\{x(t) > p_1 t + \beta \sqrt{p_2 t}\} = \Phi(-\beta) + O(t^{-\frac{1}{2}});$$

$$P\{x(t) > p_1 t + \beta \sqrt{p_2 t}\} = \Phi(-\beta) + \frac{C_3}{3!\,t^{\frac{1}{2}}} \Phi^{(3)}(-\beta)$$

(E)

$$+ \frac{C_4}{4!\,t} \Phi^{(4)}(-\beta) + \frac{10\,C_3^2}{6!\,t} \Phi^{(6)}(-\beta) + O(t^{-3/2})$$

where
$$\Phi(x) = \frac{1}{\sqrt{2\pi}} \int_{-\infty}^{x} e^{-y^2/2} dy, \quad \Phi^{(k)}(x)$$

is the kth derivative of $\Phi(x)$, $C_n = p_n/p_2^{n/2}$, and $0(t^{-f})$ means $|\text{Error in Approx.}|$ $< A/t^f$ for sufficiently large t. A is some positive constant.

These expansions require that the claim distribution satisfy various requirements which are always fulfilled in practice. Since the expansion (E) may seem mysterious, let the reader recall the Taylor series expansion for a function $f(x)$ at some value $x=a$:

$$f(x) = f(a) + f'(a)(x-a) + \frac{f''(a)}{2!}(x-a)^2 + \frac{f'''(a)}{3!}(x-a)^3 + \dots$$

$$+ \frac{f^{(n)}(a)}{n!}(x-a)^n + R_n(x, a),$$

where
$$R_n(x, a) = \int_a^x \frac{(x-t)^n}{n!} f^{(n+1)}(t)\, dt.$$

The reader may regard $f(a)$ as $\Phi(\beta)$. Furthermore, the derivatives of $\Phi(x)$ are not as complicated as one would expect. Since

$$\Phi(x) = \frac{1}{\sqrt{2\pi}} \int_{-\infty}^{x} e^{-t^2/2} dt,$$

by the Fundamental Theorem of Calculus,

$$\Phi'(x) = \frac{1}{\sqrt{2\pi}} e^{-x^2/2}.$$

It is then easy to obtain further derivatives.

$$\Phi''(x) = \frac{1}{\sqrt{2\pi}} e^{-x^2/2}(-x).$$

$$\Phi'''(x) = \frac{1}{\sqrt{2\pi}} e^{-x^2/2}(-1) + x^2 \frac{e^{-x^2/2}}{\sqrt{2\pi}}.$$

And so on.

Let us consider an outline of the proof of (E) to see why $\Phi'(\beta)$ and $\Phi''(\beta)$ do not enter the expansion, and also why the powers of t enter the coefficients in an irregular manner. We assume that the moments p_n are finite for $n \leq 5$. Consider the moment generating function of the standardized random variable

$$\frac{p_1 t - x(t)}{\sqrt{p_2 t}}.$$

$$E\left\{\exp\left[\theta \frac{p_1 t - x(t)}{\sqrt{p_2 t}}\right]\right\} = \exp\left(\frac{p_1 t \theta}{\sqrt{p_2 t}}\right) E\left\{\exp\left[\frac{-\theta}{\sqrt{p_2 t}} x(t)\right]\right\}$$

$$= \exp\left(\frac{p_1 t \theta}{\sqrt{p_2 t}}\right) \exp\left\{t\left[M\left(\frac{-\theta}{\sqrt{p_2 t}}\right) - 1\right]\right\}.$$

It proves advantageous to use characteristic functions which, among other things, involves replacing θ by $i\tau$. Thus

$$E\left\{\exp\left[i\tau \frac{p_1 t - x(t)}{\sqrt{p_2 t}}\right]\right\} = \exp\left\{t\left[M\left(\frac{-i\tau}{\sqrt{p_2 t}}\right) - 1 + \frac{p_1 i\tau}{\sqrt{p_2 t}}\right]\right\}.$$

Now $$M\left(\frac{-i\tau}{\sqrt{p_2 t}}\right) = \int_{-\infty}^{\infty} \exp\left(\frac{-i\tau}{\sqrt{p_2 t}}\right) dP(x)$$

$$= \int_{-\infty}^{\infty} \left\{1 + \frac{-i\tau}{\sqrt{p_2 t}} x + \left(\frac{-i\tau}{\sqrt{p_2 t}}\right)^2 \frac{x^2}{2!} + \ldots\right\} dP(x)$$

$$= 1 + \frac{-i\tau}{\sqrt{p_2 t}} p_1 + \left(\frac{-i\tau}{\sqrt{p_2 t}}\right)^2 \frac{p_2}{2!} + \ldots$$

and thus

$$t\left[M\left(\frac{-i\tau}{\sqrt{p_2 t}}\right) - 1 + \frac{p_1 i\tau}{\sqrt{p_2 t}}\right] = -\frac{\tau^2}{2} + t\left[\left(\frac{-i\tau}{\sqrt{p_2 t}}\right)^3 \frac{p_3}{3!} + \left(\frac{-i\tau}{\sqrt{p_2 t}}\right)^4 \frac{p_4}{4!} + O(t^{-5/2})\right].$$

The characteristic function of $aN(0,1)$ random variable is

$$\int_{-\infty}^{\infty} e^{itx} d\Phi(x) = \frac{1}{\sqrt{2\pi}} \int_{-\infty}^{\infty} \exp\left\{itx - \frac{x^2}{2}\right\} dx = e^{-t^2/2}.$$

By integration by parts,

$$\int_{-\infty}^{\infty} e^{itx} d\Phi^{(n)}(x) = (-it)^n e^{-t^2/2} \quad \text{for} \quad n = 1, 2, \ldots.$$

If we denote $\dfrac{p_1 t - x(t)}{\sqrt{p_2 t}}$ by Z,

$$E[e^{i\tau Z}] = \int_{-\infty}^{\infty} e^{i\tau z} dP(z)$$

$$= \exp\left\{-\frac{\tau^2}{2} + \frac{(-i\tau)^3 p_3}{3!\sqrt{p_2^3 t}} + \frac{(-i\tau)^4 p_4}{4! p_2^2 t} + O(t^{-3/2})\right\}.$$

But it is true that

$$\int_{-\infty}^{\infty} e^{i\tau z} d\left\{\Phi(z) + \frac{p_3}{3!\sqrt{p_2^3 t}} \Phi^{(3)}(z) + \frac{p_4}{4! p_2^2 t} \Phi^{(4)}(z) + \ldots\right\}$$

also equals

$$\exp\left\{-\frac{\tau^2}{2} + \frac{(-i\tau)^3 p_3}{3!\sqrt{p_2^3 t}} + \frac{(-i\tau)^4 p_4}{4! p_2^2 t} + + \ldots\right\}.$$

Hence by the complete equality of functions with the same Fourier transform,

$$P(\beta) = \Phi(\beta) + \frac{p_3}{3!\sqrt{p_2^3 t}} \Phi^{(3)}(\beta) + \frac{p_4}{4! p_2^2 t} \Phi^{(4)}(\beta) + \ldots.$$

The reader who studies page 32 of [21] and its references will see why the p_5 and p_6 terms may be replaced by the p_3^2 term.

We will now discuss the Esscher approximation to $F(x, t)$, as explained on pages 33–39 of [21]. The reader would profit by consulting some of the other references for alternate explanations, and further material.

The Esscher formula also uses an expansion in terms of the normal distribution and its derivatives, but it first transforms the density of the compound Poisson function so that the required value of x is translated to the portion of the real axis where the normal approximation performs best in terms of fit. We will see that this significantly decreases the error term.

Assume that we wish to calculate

$$P\left[\sum_{i=1}^{N(t)} X_i > p_1 t - Ct\right] \quad \text{for some } C.$$

An interesting C would be $C = -3\sqrt{p_2 t}$ in which case we would be seeking

$$P\left[\sum_{i=1}^{N(t)} X_i > p_1 t + 3\sqrt{p_2 t}\right].$$

Determine the unique root h of the equation

$$p_1 - \int_{-\infty}^{\infty} y e^{hy} dP(y) = C.$$

If $C = 0$, clearly $h = 0$.
If $C > 0$, then $h < 0$ and if $C < 0$, then $h > 0$.

Let
$$\beta = 1 + \int_{-\infty}^{\infty} (hy - 1) e^{hy} dP(y)$$

$$= 1 + hp(h)\bar{p}_1 - p(h)$$

where
$$p(h) = \int_{-\infty}^{\infty} e^{hy} dP(y), \quad \text{and}$$

$$\bar{p}_n = \frac{1}{p(h)} \int_{-\infty}^{\infty} y^n e^{hy} dP(y).$$

As usual, we assume that the claim distribution allows these maneuvers, which is true in all practical situations. Then if $C > 0$,

$$F(p_1 t - Ct, t) = e^{-\beta t}\left[\frac{1}{|h|\sqrt{2\pi t p(h) \bar{p}_2}} + O\left(\frac{1}{t}\right)\right]$$

and if $C < 0$, $1 - F(p_1 t - Ct, t)$ equals the same quantity. Notice that this first Esscher approximation improves the accuracy over the first normal approximation by a factor of $1/\sqrt{t}$.

The second Esscher approximation says that if $C>0$

$$F(p_1 t - Ct, t) = \frac{e^{-\beta t}}{\sqrt{2\pi}} \left[\frac{1}{u} + \frac{K_3}{u^3} + O\left(\frac{1}{u^4}\right) \right]$$

where
$$u = |h| \sqrt{tp(h)\,\bar{p}_2},$$
$$K_3 = 1 - 3k_3 + 3k_4 + 15k_6,$$
$$k_3 = h\bar{p}_3/(3!\,\bar{p}_2),$$
$$k_4 = h^2\bar{p}_4/(4!\,\bar{p}_2),$$

and
$$k_6 = k_3^2/2.$$

If $C<0$, $1-F(p_1 t - Ct, t)$ equals the same quantity. Notice that this second Esscher approximation improves the accuracy over the second normal approximation by a factor of $1/\sqrt{t}$.

The readers will find that the other presentations of the Esscher method are in terms of some auxiliary functions which have been tabulated in various journals. However, Cramér's presentation seems more straightforward and hence was followed here. Harald Bohman made a major contribution to the Esscher method in [14].

We will now consider a third approximate method which is easy to apply, and accurate. It approximates $F(x, t)$ by a function of incomplete gamma functions. It was studied first by D. K. Bartlett [4], although a much more complete presentation was made by Newton Bowers [17]. The Bowers paper was in terms of general random variables, and led to the author's approximation of the ruin function for $T = +\infty$.

For $x>0$, let the gamma density be denoted by

$$g(x) = x^{\alpha-1} e^{-x}/\Gamma(\alpha)$$

where
$$\Gamma(\alpha) = \int_0^\infty x^{\alpha-1} e^{-x} dx.$$

For $\alpha > 0$, $\Gamma(\alpha)$ is called the gamma function. It satisfies the recurrence relation

$$\Gamma(\alpha+1) = \alpha \Gamma(\alpha).$$

If α is a positive integer, $\Gamma(\alpha) = (\alpha-1)!$. For other values of α, various books have tables of the values of $\Gamma(\alpha)$. For example, the various editions of *Standard Mathematical Tables*, [36], contain values of $\Gamma(\alpha)$ for $1 \leq \alpha \leq 2$. Thus $\Gamma(4/3) = .89338$.

Consider the Laguerre polynomials

$$L_n^{(\alpha)}(x) = (-1)^n x^{1-\alpha} e^x \frac{d^n}{dx^n}(x^{n+\alpha-1} e^{-x}).$$

The first three of these are

$$L_0^{(\alpha)}(x) = 1,$$
$$L_1^{(\alpha)}(x) = x - \alpha,$$
$$L_2^{(\alpha)}(x) = x^2 - 2(\alpha+1)x + (\alpha+1)\alpha.$$

Since these polynomials are orthogonal on the positive real axis with respect to the weight function $g(x)$,

$$\frac{1}{\Gamma(\alpha)} \int_0^\infty z^{\alpha-1} e^{-z} L_n^{(\alpha)}(z) L_m^{(\alpha)}(z)\, dz = 0 \quad \text{if} \quad m \neq n.$$

If $m = n$, the integral equals $[n!\,\Gamma(\alpha+n)]/\Gamma(\alpha)$. Hence, if a given density function can be written as

$$f(x) = \frac{x^{\alpha-1} e^{-x}}{\Gamma(\alpha)} [A_0 L_0^{(\alpha)}(x) + A_1 L_1^{(\alpha)}(x) + A_2 L_2^{(\alpha)}(x) + \ldots],$$

the above orthogonality conditions allow the determination of A_n. Thus

$$\int_0^\infty f(z) L_n^{(\alpha)}(z)\, dz = \int_0^\infty \frac{z^{\alpha-1} e^{-z}}{\Gamma(\alpha)} [A_0 L_0^{(\alpha)}(z) + A_1 L_1^{(\alpha)}(z) + \ldots] L_n^{(\alpha)}(z)\, dz$$

$$= A_n \frac{n!\,\Gamma(\alpha+n)}{\Gamma(\alpha)}.$$

Thus
$$A_n = \frac{\Gamma(\alpha)}{n!\,\Gamma(\alpha+n)} \int_0^\infty f(z) L_n^{(\alpha)}(z)\, dz.$$

Now assume that we wish to approximate the density of a nonnegative valued random variable Y. It proves convenient to define a second random variable X by $X = \beta Y$, where β is chosen so that $E(X) = \text{Var}(X)$. Let $\alpha = E(X)$. Assume X has a density function $f(x)$ with a sufficient number of moments. We now determine A_n using $L_n^{(\alpha)}(x)$ for $\alpha = E(X) = \text{Var}(X)$.

$$A_0 = \int_0^\infty f(z) L_0^{(\alpha)}(z)\, dz = \int_0^\infty f(z)\, dz = 1,$$

$$A_1 = \frac{1}{\alpha} \int_0^\infty f(z) L_1^{(\alpha)}(z)\, dz = \frac{1}{\alpha} \int_0^\infty f(z)(z-\alpha)\, dz = 0,$$

$$A_2 = 0,$$

$$A_3 = \frac{\Gamma(\alpha)}{3!\,\Gamma(\alpha+3)} (\mu_3 - 2\alpha), \quad \text{where}$$

μ_n is the nth moment about the mean of X. Bowers lists A_4 and A_5 on page 128 of [17]. Hence the density of X may be approximated by a partial sum of the series

$$\frac{x^{\alpha-1} e^{-x}}{\Gamma(\alpha)} [1 + A_3 L_3^{(\alpha)}(x) + A_4 L_4^{(\alpha)}(x) + A_5 L_5^{(\alpha)}(x) + \ldots].$$

Of course this determines the desired distribution for Y since

$$P[Y \leq y] = P\left[\frac{1}{\beta} X \leq y\right] = \int_0^{\beta y} f(z)\, dz.$$

Furthermore, the formula only required the moments of X, not the exact form $f(x)$. This is ideal for the typical insurance company where the sample moments can be calculated.

Now we have seen before that if $Y = \sum_{i=1}^{N(t)} X_i$, then $E(Y) = p_1 t$, and $\text{Var}(Y) = p_2 t$. Thus $\alpha = E(X) = E(\beta Y) = \beta p_1 t$. But $\alpha = \text{Var}(X) = \text{Var}(\beta Y) = \beta^2 \text{Var}(Y) = \beta^2 p_2 t$. Therefore $\beta = p_1 t/(p_2 t)$, and $\alpha = (p_1 t)^2/(p_2 t)$. Bartlett and Bowers have shown that $E[(Y-p_1 t)^3] = p_3 t$, $E[(Y-p_1 t)^4] = p_4 t + 3 p_2^2 t^2$, and $E[(Y-p_1 t)^5] = p_5 t + 10 p_2 p_3 t^2$. These are the μ_1, μ_2, μ_3, μ_4, and μ_5 for A_3, A_4, A_5. If one substitutes the actual polynomials in the series, and integrates to obtain the distribution function, one gets

$$F(x, t) \doteq \Gamma(x, \alpha)(1 - A + B - C) + \Gamma(x, \alpha+1)(3A - 4B + 5C)$$
$$+ \Gamma(x, \alpha+2)(-3A + 6B - 10C)$$
$$+ \Gamma(x, \alpha+3)(A - 4B + 10C)$$
$$+ \Gamma(x, \alpha+4)(B - 5C) + \Gamma(x, \alpha+5) C$$

where $\Gamma(x, \alpha)$ is the incomplete gamma function

$$\Gamma(x, \alpha) = \frac{1}{\Gamma(\alpha)} \int_0^x z^{\alpha-1} e^{-z}\, dz,$$

$$A = \frac{\mu_3 - 2\alpha}{3!},$$

$$B = (\mu_4 - 12\mu_3 - 3\alpha^2 + 18\alpha)/4!,$$

and
$$C = [\mu_5 - 20\mu_4 - (10\alpha - 120)\mu_3 + 60\alpha^2 - 144\alpha]/5!.$$

One can use integration by parts to show that

$$\Gamma(x, \alpha+1) = \Gamma(x, \alpha) - \frac{x^\alpha e^{-x}}{\Gamma(\alpha+1)}.$$

Bowers uses this to derive an alternate expression for $F(x, t)$:

$$F(x, t) \doteq \Gamma(x, \alpha) - A\left[\frac{x^\alpha e^{-x}}{\Gamma(\alpha+1)} - \frac{2x^{\alpha+1} e^{-x}}{\Gamma(\alpha+2)} + \frac{x^{\alpha+2} e^{-x}}{\Gamma(\alpha+3)}\right]$$

$$+ B\left[\frac{x^\alpha e^{-x}}{\Gamma(\alpha+1)} - \frac{3x^{\alpha+1} e^{-x}}{\Gamma(\alpha+2)} + \frac{3x^{\alpha+2} e^{-x}}{\Gamma(\alpha+3)} - \frac{x^{\alpha+3} e^{-x}}{\Gamma(\alpha+4)}\right]$$

$$- C\left[\frac{x^\alpha e^{-x}}{\Gamma(\alpha+1)} - \frac{4x^{\alpha+1} e^{-x}}{\Gamma(\alpha+2)} + \frac{6x^{\alpha+2} e^{-x}}{\Gamma(\alpha+3)} - \frac{4x^{\alpha+3} e^{-x}}{\Gamma(\alpha+4)}\right.$$

$$\left. + \frac{x^{\alpha+4} e^{-x}}{\Gamma(\alpha+5)}\right].$$

Essentially Bartlett's two moment approximation was $F(x, t) \doteq \Gamma(\beta x, \alpha)$. With $\alpha = (p_1 t)^2/(p_2 t)$, we have

$$F(x, t) \doteq \Gamma(\beta x, \alpha) = \frac{1}{\Gamma(\alpha)} \int_0^{\beta x} z^{\alpha-1} e^{-z} dz.$$

The "A" term involves the third moment, the "B" term involves the fourth moment, and the "C" term increases the accuracy to fifth moments. Bowers considers two examples and in the one for which the exact values are known, his method produced very accurate answers.

In view of the fact that the gamma expansion method is easier to apply than the Esscher approximation, a very worthwhile research project would be to compare their accuracy for the same practical distribution. Apparently the "exact" values would have to be calculated by a Monte Carlo technique similar to one described in section 3.3.

Exercises

1. Verify that $E\{x(t)\} = p_1 t$ and that Std. Dev. $\{x(t)\} = \sqrt{p_2 t}$. You may wish to compare with problem 1, Chapter 12 of [22].
2. If $p_1 = 10$, $p_2 = 10$, $t = 100$, approximate $P\{x(100) > 1\,020\}$ by the Lundberg approximation. Use one of the Cramér results to approximate the error made, up to a constant A.

3. Compute $P\{x(100)>1\,020\}$ by the four term Cramér normal expansion and indicate the approximate error made, up to a constant A, Assume that $p_3=30$, $p_4=120$.

Suggestions: There are various tables of the derivatives of the normal density. One readily available source is *Handbook of Tables for Probability and Statistics*, 2nd Edition, William H. Beyer, Editor, The Chemical Rubber Company, Cleveland, 1968. The reader should observe that $\Phi'(x)=f(x)$ (the normal density), $\Phi''(x)$ is the derivative of the density, etc. Furthermore, $\Phi^{vi}(x)=f^v(x)=-xf^{iv}(x)-4f^{iii}(x)$. Although the tables only use $x \geqslant 0$, negative values of x can be handled since $f(-x)=f(x)$, $f'(-x)=-f'(x)$, $f''(-x)=f''(x)$, $f'''(-x)=-f'''(x)$, $f^{iv}(-x)=f^{iv}(x)$, $f^v(-x)=-f^v(x)$.

4. In the equation

$$p_1 - \int_{-\infty}^{\infty} y e^{hy} dP(y) = C$$

used in the Esscher approximation, let

$$P(y) = \begin{cases} 1-e^{-y}, & y \geqslant 0 \\ 0, & y < 0 \end{cases}$$

and solve for h.

5. Verify the expression for $L_2^{(\alpha)}(x)$. Evaluate A_2 in the gamma function expansion.

6. If $Y=\sum_{i=1}^{N(t)} X_i$, prove that $E[(Y-p_1 t)^3]=p_3 t$, and that

$$E[(Y-p_1 t)^4]=p_4 t+3p_2^2 t^2.$$

7. Assume that $P(x)=1-e^{-x}$, $x \geqslant 0$ and that $t=16$. Using only the first term in the gamma expansion, compute $F(x,t)$ for $x=0, 4, 20$, and 40.

3.2. A CONVOLUTION FORMULA FOR $\psi(u)$

The following theorem appears in [8]. It is based on a paper by L. Takács [39] which concerns stochastic processes with stationary independent increments. This theorem gives an expression for $\psi(u)$ in terms of the convolutions of a distribution related to the claim distribution $P(x)$.

THEOREM. Assume that the common distribution function $P(x)$ is such that $\lim_{x \to \infty} x[1-P(x)]=0$.

Let $H_0^*(x)=1$ if $x \geqslant 0$ and 0 if $x<0$. Let

$$H^*(x) = \frac{1}{p_1}\int_0^x [1-P(y)]\,dy$$

for $x \geqslant 0$ and 0 for $x<0$;

$$H_1^*(x) = H^*(x); \quad H_n^*(x) = \int_0^x H_{n-1}^*(x-z)\,dH^*(z)$$

for $x \geqslant 0$ and 0 for $x<0$; $n \geqslant 1$. Then for $u \geqslant 0$,

$$\psi(u) = 1 - \frac{\lambda}{p_1+\lambda}\sum_{n=0}^{\infty}\left(\frac{p_1}{p_1+\lambda}\right)^n H_n^*(u).$$

Proof. First observe that $H^*(x)$ is a distribution function. This is true because $H^*(x)$ is nondecreasing for $-\infty < x < \infty$, continuous from the right for $-\infty < x < \infty$, $\lim_{x \to -\infty} H^*(x) = 0$, and $\lim_{x \to +\infty} H^*(x) = 1$. The fact that $\lim_{x \to +\infty} H^*(x) = 1$ takes some explanation. By the integration by parts formula for Stieltjes integrals,

$$H^*(x) = \frac{1}{p_1}[1-P(x)]x - \frac{1}{p_1}\int_0^x y\,d[1-P(y)].$$

By properties of Stieltjes integrals,

$$H^*(x) = \frac{1}{p_1}[1-P(x)]x + \frac{1}{p_1}\int_0^x y\,dP(y).$$

Using our assumption on $P(x)$, the above $\to 1$ as $x \to \infty$. Next note that

$$\alpha p_1 \int_0^{\infty} e^{-\alpha x}\,dH^*(x) = \left[1 - \int_0^{\infty} e^{-\alpha x}\,dP(x)\right], \quad \alpha \geqslant 0.$$

Substituting this in Corollary 4 gives

$$\psi(u) = 1 - \lambda I_\alpha \left\{ \frac{1}{\alpha(p_1+\lambda)\left[1-\dfrac{p_1}{p_1+\lambda}\displaystyle\int_0^{\infty} e^{-\alpha x}\,dH^*(x)\right]} \right\}.$$

Now use the power series for $(1-z)^{-1}$, namely, $(1-z)^{-1} = 1+z+z^2+z^3+\ldots$, $|z|<1$, for the quantity

$$1\left/\left[1 - \frac{p_1}{p_1+\lambda}\int_0^{\infty} e^{-\alpha x}\,dH^*(x)\right]\right..$$

By properties of the Laplace transform of a convolution,

$$\left[\int_0^\infty e^{-\alpha x} dH^*(x)\right]^k = \int_0^\infty e^{-\alpha x} dH_k^*(x).$$

(This is similar to the equality of the moment generating function of the sum of k independent, identically distributed random variables with the kth power of the common moment generating function.)

Combining these two facts,

$$\psi(u) = 1 - \frac{\lambda}{p_1 + \lambda} I_\alpha \left\{ \left[1 + \frac{p_1}{p_1 + \lambda} \int_0^\infty e^{-\alpha x} dH^*(x) \right. \right.$$
$$\left. \left. + \left(\frac{p_1}{p_1 + \lambda}\right)^2 \int_0^\infty e^{-\alpha x} dH_2^*(x) + \ldots \right] \Big/ \alpha \right\}.$$

By integration by parts, and the fact that $H_k^*(0) = 0$ for $k \geq 1$,

$$\psi(u) = 1 - \frac{\lambda}{p_1 + \lambda} I_\alpha \left[\frac{1}{\alpha} + \frac{p_1}{p_1 + \lambda} \int_0^\infty e^{-\alpha x} H^*(x) \, dx \right.$$
$$\left. + \left(\frac{p_1}{p_1 + \lambda}\right)^2 \int_0^\infty e^{-\alpha x} H_2^*(x) \, dx + \ldots \right].$$

Since

$$I_\alpha \left[\int_0^\infty e^{-\alpha x} H_k^*(x) \, dx \right] = H_k^*(x),$$

term by term inversion of the above expression gives the formula of the theorem.

Several remarks can be made about this expression for $\psi(u)$. First, it is *remarkably similar* to the traditional formula for $F(x, t)$:

$$F(x, t) = \sum_{n=0}^\infty \frac{t^n e^{-t}}{n!} P_n^*(x),$$

where $P_n^*(x)$ is the nth convolution of the claim distribution. Secondly, the coefficients of the H_n^*'s are the probabilities of a geometric distribution. Our assumption that $\lim_{x \to \infty} x[1 - P(x)] = 0$ is a very mild restriction on $P(x)$. Indeed, the reader is challenged to find a $P(x)$ which does not fulfill this requirement.

Several properties of $H_n^*(u)$ are helpful:

(1) $H_n^*(u)$ is continuous for all u, $n \geqslant 1$.

(2) For fixed u, $\lim_{n \to \infty} H_n^*(u) = 0$.

(3) For fixed n, $\lim_{u \to \infty} H_n^*(u) = 1$.

If $\lambda = kp_1$, $0 < k < 1$, and there exists an integer $n(p_1, u)$ (which depends on p_1 and u) for which $H_j^*(u) = 1$, $1 \leqslant j \leqslant n(p_1, u)$ then

$$\psi(u) \doteq \left(\frac{1}{1+k}\right)^{n+1} \text{ with error } \leqslant \left(\frac{1}{1+k}\right)^{n+1}.$$

This approximation overstates the true probability of ruin. For small, but meaningful, values of $\psi(u)$, this expression with $p_1 = 1$ gave results which closely agreed with the table in [7] for $\psi(u)$ for $P(x) = 0$, $x < 1$ and $P(x) = 1$, $x \geqslant 1$.

As an example, consider the two distributions:

$$P_1(x) = \begin{cases} 0, & x < 10 \\ 1, & x \geqslant 10 \end{cases};$$

$$P_M(x) = \begin{cases} 0, & x < 10M \\ 1, & x \geqslant 10M \end{cases}.$$

As usual, the unit of money is $1 000.00.

For $P_1(x)$, $H_j^*(u) = 1$ for $j \leqslant .1u$. For $P_M(x)$, $H_j^*(u) = 1$ for $j \leqslant .1u/M$. Now assume that $\lambda = .3p_1$. Then for $P_1(x)$ and $u = 100$,

$$\psi(100) = (1/1.3)^{11} \doteq .0556 \text{ with error } \leqslant .0556.$$

whereas for $P_M(x)$, $\psi(M \times 10^2) \doteq .0556$ with error $\leqslant .0556$.

We reach the conclusion that increasing the average claim by a factor of M only requires a like multiplication of the initial capital to preserve the same probability of ruin.

Exercises

1. Let $P(x) = 1 - e^{-Ax}$, $x \geqslant 0$, $A > 0$. Compute $F(x, t)$ by the convolution method.

2. Let $P(x) = \begin{cases} 0, & x < p_1 \\ 1, & x \geqslant p_1 \end{cases}$.

   Compute $F(x, t)$ by the convolution method.

3. For $P(x)$ as in problem 1, compute $\psi(u)$ by the convolution method.

3.3. A Monte Carlo approach to $\psi(u, T)$

The following approximation of $\psi(u, T)$ is based on a paper by Hilary Seal [33].

So far, we have considered the total claims as a random sum of random variables, $\sum_{i=1}^{N(t)} X_i$ where $N(t)$ has a Poisson distribution with mean t. But there is an alternate way of viewing this which has certain theoretical advantages and marked advantages if one is going to use simulation to approximate $\psi(u, T)$. The lengths of the times between claims are random variables, each of which may be assumed to have as its probability density function $t^{-1} e^{-\tau/t}$ which means that the average number of claims in unit times is $1/t$.(¹) In other words, if L is the random variable representing the time until the first claim, then

$$P[L > x] = \int_x^\infty \frac{1}{t} e^{-y/t} dy = e^{-x/t}.$$

Similarly, if T represents the random time between the ith and the $(i+1)$st claim, then $P[T > x] = e^{-x/t}$. As far as actuaries are concerned, this idea may be traced to the 1903 doctoral thesis of Filip Lundberg. Looking back at section 3.0, we see that

$$P[N(t+s) - N(t) = 0] = e^{-s}$$

where we assumed the rate of occurrence per unit time was 1.

At time t, the risk reserve $U(t)$ is given by

$$U(t) = u + (p_1 + \lambda) t - \sum_{i=1}^{N(t)} X_i.$$

We will examine $U(t)$ at the claim times t_1, $t_1 + t_2$, ..., $\sum_{i=1}^n t_i$ where $\sum_{i=1}^n t_i \leqslant T$. If $U(t) \geqslant 0$ for each claim time this particular simulated company is not ruined during the period $[0, T]$. If n such company histories are simulated, and k are ruined in the time $[0, T]$, then the estimate of $\psi(u, T)$ is

$$\hat{\psi}(u, T) = k/n$$

with an estimated standard error of

$$\left[\frac{k/n(1 - k/n)}{n} \right]^{\frac{1}{2}}.$$

(¹) Hence the average number of claims in one operational time unit is one.

The reader is reminded that if an experiment is performed n independent times, and if the random outcomes $X_1, X_2, ..., X_n$ each equal 1 (success) with probability p, and 0 (failure) with probability q, then the maximum likelihood estimator for p is $\sum_{i=1}^{n} X_i/n = \bar{X}$ with variance $\sigma^2(\bar{X}) = pq/n$. In this case, the variance was estimated by replacing p by k/n. A "1" corresponded to a company being ruined.

Before discussing the simulation of $\psi(u, T)$, some remarks should be made about the random numbers. Let us assume that P[Person aged x dies within 1 year] $= q_x$. If we use a supply of random numbers distributed uniformly on $[0, 1]$, then P[Random number $\leq q_x$] $= q_x$. Therefore we may simulate the experiment of observing a person aged x for one year by drawing a random number and recording his death if the random number is $\leq q_x$. That is, the two events

$$A = \{\text{Person aged } x \text{ dies within 1 year}\} \quad \text{and}$$

$$B = \{\text{Random number } \leq q_x\} \quad \text{have equal probabilities.}$$

It should be obvious to the reader that supplies of random numbers distributed according to a normal density, or a negative exponential density or a Poisson law are as readily available as those which are uniformly distributed.

Perhaps it is worth repeating the explanation in different words, following Seal's paper rather closely. The simulation for one company consists of four steps.

1. Select a suitable claim distribution $P(x)$, and values for u, λ, and T.
2. Compute a random negative exponential variate, t_1, from $P[T \leq t] = 1 - e^{-t}$, $t \geq 0$, and a variate from $P(x)$, x_1.
3. Calculate $U(t_1) = u + (p_1 + \lambda) t_1 - x_1$. If $U(t_1) < 0$, ruin has occurred and we start a new company.
4. If $U(t_1) \geq 0$, compute another pair of variates, t_2 and x_2, calculate $U(t_1 + t_2) = u + (p_1 + \lambda)(t_1 + t_2) - (x_1 + x_2)$, and proceed as with $U(t_1)$.
5. Continue to compute pairs of variates (t_j, x_j) for $j = 3, 4, ...$, and calculate the corresponding U's. If n is the smallest integer such that

$$\sum_{j=1}^{n} t_j \leq T < \sum_{j=1}^{n+1} t_j \quad \text{and} \quad U\left(\sum_{j=1}^{n} t_j\right) \geq 0,$$

one says that the company has not been ruined in the period $[0, T]$. By running a large number of simulations, one can make the standard error E as small as possible. Although one should recognize there is always a positive probability ε that $\psi(u, T)$ is outside the band

$$[\hat{\psi}(u, T) - kE, \hat{\psi}(u, T) + kE],$$

by increasing k and keeping E small by increasing n, one can force ε to be as small as one likes. Thus this method is as accurate as one's budget will allow.

However, the method is expensive. Thus for the claim distribution $P(x) = 1 - e^{-x}$, $x \geqslant 0$, and 60 000 trials of each of nine combinations of u and λ, the aggregate computer time of a fast and expensive model was one hour and 45 minutes. This is part of the motivation for the next section.

Simulation of $\psi(u, T)$ possesses several advantages which should be mentioned. The distribution of interclaim periods can be quite general, and the distribution of claims can be made to depend on the times of claims.

Exercise. Derive a simulation program for $F(x, t)$.

3.4. A MOMENT APPROACH TO $\psi(u, T)$

Newton Bowers and the author have derived an approximation for $\psi(u, T)$ in [11] and [12]. Here is a summary of it.

Let $X_2, X_2, \ldots$ be a sequence of independent, identically distributed random variables with $P[X \leqslant 0] = 0$, and such that $E(X^4) < \infty$. Assume that $[N(t), t \geqslant 0]$ is a Poisson stochastic process independent of the X_i's with $E[N(t)] = t$. For $\lambda \geqslant 0$, let

$$Z_T = \max_{0 \leqslant t \leqslant T} \left[\sum_{i=1}^{N(t)} X_i - t(p_1 + \lambda) \right].$$

Expressions are derived for $E(Z_T)$, $E(Z_T^2)$, and $P(Z_T = 0)$ by first approximating the solution of an integral equation contained in [7]. These results agree, as $T \to \infty$, with earlier results for $E(Z_\infty)$, $E(Z_\infty^2)$, and $P[Z_\infty = 0]$ [9]. The finite T results are used to derive an approximation for the finite-time ruin function $\psi(u, T) = P[Z_T > u]$ for $u \geqslant 0$. The accuracy of the approximations is discussed. It should be mentioned that the time moments

$$\int_0^\infty T^k d_T \psi(u, T), \; k = 1, 2$$

where considered by C. O. Segerdahl, who used them to approximate (by a normal distribution) the time to ruin, given that it is certain, for large values of u, and T.

The formula for $E[Z_T^2]$ allowed the proof of the

THEOREM. $\psi(u, T) \leqslant p_2 T/u^2$ for $\lambda \geqslant 0$ and $u > 0$. The reader will quickly observe that there are some combinations of p_2, T, and u for which the upper bound exceeds one, and hence is useless. However, for many other combinations it provides a very fast way of roughly checking the effects of changing p_2, T and u.

Exercise. Consider the two distributions

$$P_1(x) = \begin{cases} 0, & x < 1 \\ 1, & x \geqslant 1 \end{cases} \qquad P_{10}(x) = \begin{cases} 0, & x < 10 \\ 1, & x \geqslant 10 \end{cases}$$

If we allow $T = 10$, and $\lambda = 0$, what respective values of u will hold $\psi(u, T) \leqslant .001$ in both cases?

3.5. INVERTING TRANSFORMS TO OBTAIN $\psi(u, T)$ AND $\psi(u)$

Although the simulation method of obtaining approximations to $\psi(u, T)$ is easy to understand, it is an expensive method. The approximation of $\psi(u, T)$ by the use of the moments of Z_T and the incomplete gamma distribution is not accurate enough, in some cases. A method will now be explained which avoids these disadvantages. For the sake of simplicity, it will only be explained in one of its special cases. We will assume that the claim interoccurrence times are distributed by the negative exponential distribution, and the claim distribution can be expressed by a linear combination of five negative exponential distributions. Thus, if T is a random variable representing the time between claims,

$$P[T \leqslant t] = \begin{cases} 1 - e^{-t}, & t \geqslant 0 \\ 0, & t < 0 \end{cases}$$

and the claim distribution

$$P(y) = 1 - \sum_{n=1}^{5} a_n e^{-b_n y}$$

where $\sum_{n=1}^{5} a_n = 1$, and $b_n > 0$, $n = 1, 2, 3, 4, 5$.

The method is more general with respect to both the distributions of claim times and amounts, and the reader is invited to consult references [40],

[41], and [44] in which Olof Thorin has derived the theory and Nils Wikstad has shown how the theory can be implemented.

For complex z, with Re $(z) \leq 0$, let $\bar{\psi}(u, z) = \int_0^\infty e^{zT} d_T \psi(u, T)$. Let us define $\bar{\psi}(u, z) = 0$ for $u < 0$, and let $\bar{\varphi}(s, z) = 1 - \int_{0-}^\infty e^{su} d_u \bar{\psi}(u, z)$. To obtain $\psi(u, T)$, first invert $1 - \bar{\varphi}(s, z)$ to obtain $\bar{\psi}(u, z)$ and then invert $\bar{\psi}(u, z)$. Theoretically, these inversions would be done by the Lévy inversion formula. A numerical method is Bohman's C-method explained in [15]. The first inversion must be done for a large number of z-values in order to get a sufficient basis for the second inversion. This double inversion would be expensive. Thorin avoids the first inversion by his assumption on $P(y)$. If $p(s)$ is the Fourier transform of $P(y)$,

$$p(s) = \int_0^\infty e^{sy} dP(y)$$

$$= \sum_{n=1}^5 \int_0^\infty e^{sy} a_n b_n e^{-b_n y} dy$$

$$= \sum_{n=1}^5 \frac{a_n b_n}{b_n - s} = \sum_{n=1}^5 \frac{a_n}{1 - s/b_n}$$

for $|s| < \min \{b_1, b_2, b_3, b_4, b_5\}$. Let $S_{2n}(z)$, $n = 1, 2, 3, 4, 5$ be the zeros of $p(s) = 1 + (p_1 + \lambda)s - z$ for Re $(s) > 0$.

This is an equation of the sixth degree in s, but it is only necessary to compute five roots as one root is in the left half plane. Thorin then shows that

$$\bar{\psi}(u, z) = \sum_{n=1}^5 g_n(z) e^{-u S_{2n}(z)}, \quad u \geq 0$$

where

$$g_j(z) = \frac{\prod_{n=1}^5 (1 - S_{2j}(z)/b_n)}{\prod_{\substack{n=1 \\ n \neq j}}^5 (1 - S_{2j}(z)/S_{2n}(z))}$$

for $j = 1, 2, 3, 4, 5$. The $S_{2n}(z)$ are roots of polynomials with complex coefficients because of the appearance of z. Luckily, computer programs exist to perform such root extraction, for example in the IBM SSP (PL1) library. One then obtains $\psi(u, T)$ by a numerical inversion, e.g. according to the method proposed by R. Piessens in BIT, Vol. 9 (1969), pp. 351–361.

Tables so computed will be found in [41] and [44]. These references also give more detail about the numerical methods of finding the complex roots, and performing the numerical inversions.

Let $\Phi(t)$ be the characteristic function of the claim distribution. Assume that the mean value of the claim distribution is 1, i.e. $\Phi'(0)=i$. In [16] H. Bohman shows that the characteristic function $f(t)$ of $\psi(u)$ is given by the equation

$$f(t) = \frac{\lambda}{1+(1+\lambda)it-\Phi(t)} - \frac{1}{it}.$$

The Fourier inversion formula theoretically yields $\psi(u)$ as follows:

$$\psi(u) = \frac{1}{2\pi}\int_{-\infty}^{+\infty} f(t)\,e^{-itu}dt.$$

Bohman derives the following approximation:

$$\psi(u) \doteq \frac{\delta}{2\pi}\sum_{k=-\infty}^{\infty} f(\delta k)\,e^{-iu\delta k} = \frac{\delta f(0)}{2\pi} - \sum_{\substack{k=-\infty \\ k\neq 0}}^{\infty} \frac{e^{iu\delta k}}{2\pi i k}$$

$$+ \frac{\lambda\delta}{2\pi}\sum_{\substack{k=-\infty \\ k\neq 0}}^{\infty} \frac{e^{-iu\delta k}}{1+(1+\lambda)i\delta k-\Phi(\delta k)}$$

where δ is chosen so small that practically all mass of probability is situated in the interval $[-\pi/\delta,\ \pi/\delta]$. The first Fourier series equals $.5-\delta u/(2\pi)$. The second Fourier series is summed according to a modified Fejér method, given in [15]. These facts combine to give

$$\psi(u) \doteq \frac{\delta f(0)+\pi-\delta k}{2\pi} + \frac{\delta\lambda}{2\pi}\sum_{\substack{k=1-N \\ k\neq 0}}^{N-1} \frac{C(k/N)\,e^{-i\delta ku}}{1+(1+\lambda)i\delta k-\Phi(\delta k)}$$

where $C(t)=(1-|t|)\cos\pi t+|(\sin\pi t)/\pi|$ and N is sufficiently large. The trigonometric sum is calculated according to the Fast Fourier Transform method. A program written in FORTRAN IV is given and an example is worked out numerically. For that example, $N=1\,024$.

3.6. Net stop-loss premiums

Suppose that as a small insurance company we wish to buy a policy which will cover our aggregate losses above some amount. Or what is more to the point, assume that we are a reinsurance company or department which is selling such stop-loss coverage. What should we charge as a net premium? Let us assume that we are covering all losses above a percentage u of the expected total claims $p_1 t$. Then our net charge would be $\pi(utp_1) = \int_{utp_1}^{\infty} (x - utp_1) d_x F(x, t)$, as Kahn [28] notes by the equivalence principle. This may be interpreted as the mean of the random variable X corresponding to total claims above utp_1. Ammeter [2] suggested that a logical loading for the net premium would be $\sigma(X)$ where

$$\sigma^2(X) = \int_{utp_1}^{\infty} (x - utp_1)^2 d_x F(x, t) - [\pi(utp_1)]^2.$$

There is at least one claim distribution for which an exact expression can be given for $\pi(utp_1)$. Assume that $P(x) = 1 - e^{-Ax}$, $x \geq 0$, $A > 0$. Then $p_1 = 1/A$. We have seen earlier that

$$F(x, t) = e^{-t} \left[1 + \int_0^x e^{-wA}(tA) \sum_{y=0}^{\infty} \frac{(tAw)^y}{(y+1)! \, y!} dw \right].$$

Let u be a percentage of t/A. Then

$$\pi(ut/A) = \int_{ut/A}^{\infty} (x - ut/A) e^{-t - Ax}(tA) \sum_{y=0}^{\infty} \frac{(tAx)^y}{(y+1)! \, y!} dx$$

$$= \frac{t e^{-t(1+u)}}{A} \sum_{y=0}^{\infty} \frac{t^y}{y!} \left[\frac{(ut)^{y+1}}{(y+1)!} + \left(1 - \frac{ut}{y+1}\right) \sum_{k=0}^{y} \frac{(ut)^k}{k!} \right].$$

This uses the fact that

$$\frac{1}{y!} \int_{ut}^{\infty} w^y e^{-w} dw = 1 - \Gamma(ut, y+1) = e^{-ut} \sum_{k=0}^{y} \frac{(ut)^k}{k!}.$$

$\Gamma(ut, k+1)$ is the incomplete gamma function.

This expression was approximately calculated for various u's, t's and A's and appears as follows on page 194 of [8].

The mathematical error analysis of the series for $\pi(ut/A)$ is difficult, but partial sums using thirty-one, thirty-two, thirty-three, thirty-four, and thirty-five terms indicated that the remaining terms of the series were dominated by geometric series with various common ratios. The truncation errors were then less than $a_{35} \sum_{k=1}^{\infty} r^k = a_{35} r/(1-r)$, and several are indicated in parentheses under the premiums. The factor $[1 - ut/(y+1)]$ in the formula is negative for the preliminary terms, but convergence is quite rapid after it turns positive. It is curious that for $u = 1.30$ and 1.40, the net stop-loss premiums decrease as t increases. It would be interesting to

Net stop loss premiums ([1])

t	u				
	1.00 ($)	1.10 ($)	1.20 ($)	1.30 ($)	1.40 ($)
$A = .4$ (Average claim = $2 500.00)					
16	5 620	3 978	2 734	1 827	1 187
18	5 962	4 120	2 753	1 779	1 114
20	6 278	4 240	2 754	1 722	1 038
$A = .1$ (Average claim = $10 000.00)					
16	22 478	15 910	10 936	7 306	4 750
18	23 847 (6)[a]	16 482	11 010	7 116	4 455
20	25 111 (53)[a]	16,958 (47)[a]	11 015	6 888	4 151

[a] Truncation error.
([1]) Reprinted with the permission of the editors of *Transactions, Society of Actuaries*.

make a mathematical analysis of this. In the meantime, it can be observed that this phenomenon is consistent with Bartlett's Table 1 ([4], p. 445). Melvin McFall extended the previous table to also include $t = 25(5)50$, $u = 1.5(.1)2.0$. His net premiums used the first eighty-five terms of the series, and had smaller truncation errors. McFall's paper appears in [29].

For almost every other claim distribution, approximations must be made to $F(x, t)$. Kahn [28] considers one example in detail for which he used the Esscher approximation to calculate $\pi(utp_1)$. Bowers [17] used the gamma function approximation to derive an approximation to $\pi(utp_1)$ which he illustrated with an example from Bartlett [4]. He also presented an interesting table which showed the net premium as a percentage of μ (expected total claims) for stop-loss coverage for excess claims over $\mu + 2\sigma^2/\mu$ where σ^2 is the variance of total claims. In another paper, [19], he derived the following upper bound for the stop-loss net premium. If

$$z = \mu + K\sigma, \pi(z) \leq \frac{\sigma}{2} \frac{1}{K + \sqrt{1 + K^2}}.$$

He gives tables comparing the upper bound with the net premium calculated under various distribution assumptions about $\sum_{i=1}^{N(t)} X_i$, and also comparing the upper bound with some premiums prepared as part of the Bohman-Esscher report.

3.7. A MEASURE FOR THE COLLECTIVE RISK PROCESS

Let $\mathcal{D}_0[0, \infty)$ be the space of functions $\{w(t): 0 \leq t < \infty\}$ such that if $w \in \mathcal{D}_0$, $w(0) = 0$ and for $t > 0$ w is determined by two infinite vectors

$$(t_1, t_1 + t_2, t_1 + t_2 + t_3, \ldots)$$

$$(a_1, a_2, a_3, \ldots)$$

where $t_i > 0$ for $i = 1, 2, \ldots$

$-\infty < a_i < \infty,$ for $i = 1, 2, \ldots$

$a_{i+1} - a_i \neq 0, \quad i = 1, 2, \ldots$

and $$\sum_{i=1}^{\infty} t_i = +\infty.$$

The function w is determined as follows:

$$w(t) = 0, \quad 0 \leq t < t_1$$

$$w(t) = a_i, \quad \sum_{j=1}^{i} t_j \leq t < \sum_{j=1}^{i+1} t_j, \quad i = 1, 2, \ldots.$$

Such a w function has the appearance

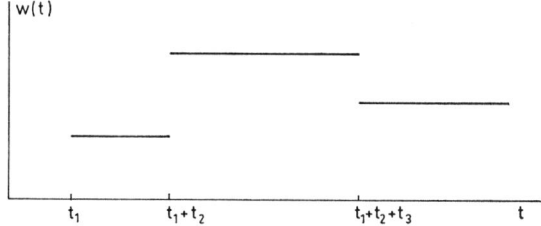

The condition $\sum_{i=1}^{\infty} t_i = +\infty$ allows at most a finite number of discontinuities in any finite interval. For if for some $T > 0$, for every integer n, $\sum_{i=1}^{n} t_i \leq T$, we would have a contradiction. This condition is suggested on p. 23 of [21] by H. Cramér. Much of this section is based on pages 23–28 and 50–52 of [21].

We will now build a measure on $\mathcal{D}_0[0, \infty)$. We will need to assign a probability measure to sets of the form:

$$\{w(s_1) \leq x_1, w(s_2) \leq x_2, \ldots, w(s_n) \leq x_n\}.$$

Let $X_1, X_2, \ldots$ be a sequence of independent, identically distributed random variables with common distribution $F(x)$. Let $T_1, T_2, \ldots$ be a

79

sequence of independent, identically distributed random variables with common distribution $P[T \leq t] = 1 - e^{-t}$, $t > 0$, and independent of X_i's. For any fixed t, the probability that the two relations

$$T_1 + T_2 + \ldots + T_n \leq t$$
$$T_1 + T_2 + \ldots + T_{n+1} > t$$

are jointly satisfied is $t^n e^{-t}/n!$. Let $\{N(t), 0 \leq t < \infty\}$ be a Poisson stochastic process independent of the X_i's with $E\{N(t)\} = t$.

Let $x(t) = \sum_{i=1}^{N(t)} X_i$, $0 \leq t < \infty$ with the agreement that in this definition and in the following, $\sum_{i=a}^{b} X_i = 0$ if $b < a$. By the assumptions made,

$$F(x, t) = P[x(t) \leq x]$$
$$= \sum_{n=0}^{\infty} \frac{t^n e^{-t}}{n!} P^{n*}(x)$$

where
$$P^{0*}(x) = \begin{cases} 0, & x < 0 \\ 1, & x \geq 0 \end{cases}$$

$$P^{1*}(x) = F(x)$$

and for $n > 1$
$$P^{n*}(x) = \int_0^{\infty} P^{(n-1)*}(x - z) \, dF(z).$$

We will now show that the increments are independent and stationary. Let $0 \leq t_0 < t_1 < \ldots < t_n \leq T$ and assume $u_1, u_2, \ldots, u_n$ are real numbers. Let $\varphi_X(u) = E\{e^{iuX}\}$.

$$E\left\{\exp\left[i \sum_{k=1}^{n} u_k [x(t_k) - x(t_{k-1})]\right]\right\}$$

$$= E\left\{\exp\left[i \sum_{k=1}^{n} u_k \sum_{j=N(t_{k-1})+1}^{N(t_k)} X_j\right]\right\}$$

$$= \sum_{m_0 \leq m_1 \leq \ldots \leq m_n} E\left\{\exp\left[i \sum_{k=1}^{n} u_k \sum_{j=N(t_{k-1})+1}^{N(t_k)} X_j\right] \middle| N(t_0) = m_0, \right.$$
$$\left. N(t_1) = m_1, \ldots, N(t_n) = m_n\right\} P[N(t_0) = m_0, \ldots, N(t_n) = m_n]$$

$$= \sum_{m_0 \leq m_1 \leq \ldots \leq m_n} E\left\{\exp\left[i \sum_{k=1}^{n} u_k \sum_{j=m_{k-1}+1}^{m_k} X_j\right]\right\}$$
$$\times P[N(t_0) = m_0, \ldots, N(t_n) = m_n]$$

$$= \sum_{m_0 \leq m_1 \leq \ldots \leq m_n} \prod_{k=1}^{n} [\varphi_X(u_k)]^{m_k - m_{k-1}} P[N(t_0) = m_0,$$

$$N(t_1) - N(t_0) = m_1 - m_0, \ldots, N(t_n) - N(t_{n-1}) = m_n - m_{n-1}]$$

$$= \sum_{m_0 \leq m_1 \leq \ldots \leq m_n} \prod_{k=1}^{n} [\varphi_X(u_k)]^{m_k - m_{k-1}}$$

$$\frac{e^{-t_0} t_0^{m_0}}{m_0!} \prod_{j=1}^{n} \frac{e^{-(t_j - t_{j-1})} [t_j - t_{j-1}]^{m_j - m_{j-1}}}{(m_j - m_{j-1})!}$$

$$= \sum_{m_0 \leq m_1 \leq \ldots \leq m_n} \left\{ e^{-t_n} \prod_{k=0}^{n} \frac{[(t_k - t_{k-1}) \varphi_X(u_k)]^{m_k - m_{k-1}}}{(m_k - m_{k-1})!} \right\}$$

where we let $t_{-1} = 0$, $u_0 = 0$, $m_{-1} = 0$. Now change variables: $m_k - m_{k-1} = 0_k$, $k = 0, 1, \ldots, n$. Then the previous

$$= \sum_{0_n = 0}^{\infty} \ldots \sum_{0_1 = 0}^{\infty} \sum_{0_0 = 0}^{\infty} e^{-t_n} \prod_{k=0}^{n} \frac{[(t_k - t_{k-1}) \varphi_X(u_k)]^{0_k}}{0_k!}$$

$$= e^{-t_n} \prod_{k=0}^{n} \exp[(t_k - t_{k-1}) \varphi_X(u_k)]$$

$$= \prod_{k=1}^{n} \exp\{(t_k - t_{k-1})[\varphi_X(u_k) - 1]\}.$$

Therefore the increments are independent and stationary. To specify the joint probability law of the n random variables $x(s_1), x(s_2), \ldots, x(s_n)$ for $s_1 < s_2 < \ldots < s_n$ we will give the joint characteristic function $\varphi_{x(s_1), \ldots, x(s_n)}(u_1, \ldots, u_n)$.

$$\varphi_{x(s_1), \ldots, x(s_n)}(u_1, \ldots, u_n)$$
$$= E[\exp[i(u_1 x(s_1) + \ldots + u_n x(s_n))]]$$

where $u_1, u_2, \ldots, u_n$ are real numbers. In this instance, we will use the notation $\overline{a - b} = (a - b)$. Then

$$E\{\exp[i(u_1 x(s_1) + \ldots + u_n x(s_n))]\}$$

$$= E\left\{\exp\left[i\left(\sum_{i=1}^{n} u_i x(s_1) + \sum_{i=2}^{n} u_i \overline{x(s_2) - x(s_1)} + \sum_{i=3}^{n} u_i \overline{x(s_3) - x(s_2)}\right.\right.\right.$$
$$\left.\left.\left. + \ldots + (u_{u-1} + u_n) \overline{x(s_{n-1}) - x(s_{n-2})} + u_n \overline{x(s_n) - x(s_{n-1})}\right)\right]\right\}$$

$$= \varphi_{x(s_1)}\left(\sum_{i=1}^{n} u_i\right) \prod_{k=2}^{n} \varphi_{x(s_k) - x(s_{k-1})}\left(\sum_{i=k}^{n} u_i\right)$$

by independent increments. For $k = 1, 2, \ldots$ (with $s_0 = 0$)

$$\varphi_{x(s_k)-x(s_{k-1})}(u) = E\{\exp[iu\overline{x(s_k)-x(s_{k-1})}]\}$$

$$= \int_0^\infty e^{iuy} d_y P[x(s_k)-x(s_{k-1}) \leqslant y]$$

$$= \int_0^\infty e^{iuy} d_y P\left[\sum_{i=1}^{N(s_k-s_{k-1})} X_i \leqslant y\right]$$

$$= \sum_{n=0}^\infty \frac{(s_k-s_{k-1})^n e^{-(s_k-s_{k-1})}}{n!} \int_0^\infty e^{iuy} dP^{n*}(y)$$

$$= \sum_{n=0}^\infty \frac{(s_k-s_{k-1})^n e^{-(s_k-s_{k-1})}}{n!} [p(u)]^n$$

where $\quad p(u) = \int_0^\infty e^{iuy} dF(y) = \exp\{(s_k-s_{k-1})[p(u)-1]\}.$

Thus

$$\varphi_{x(s_1),\ldots,x(s_n)}(u_1,\ldots,u_n) = \prod_{k=1}^n \exp\left\{(s_k-s_{k-1})\left[p\left(\sum_{i=k}^n u_i\right)-1\right]\right\}.$$

Let $\mu_{t_1,t_2,\ldots,t_n}$ be the measure determined by the requirement that

$$P_0[w(t_1) \leqslant x_1, w(t_2) \leqslant x_2, \ldots, w(t_n) \leqslant x_n]$$
$$= P[x(t_1) \leqslant x_1, x(t_2) \leqslant x_2, \ldots, x(t_n) \leqslant x_n].$$

This set function satisfies Kolmogorov's symmetry and consistency conditions, and hence determines a unique probability distribution on the Borel sets of $\mathcal{D}_0[0, \infty)$.

Let us now consider defining a measure for the sample functions of the process

$$y(t) = x(t) - kt, \quad 0 \leqslant t < \infty$$

for $k > 0$. Such a path would have the appearance

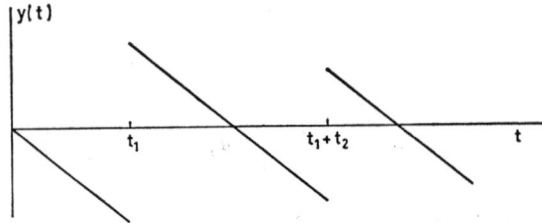

For practical purposes, $k = p_1 + \lambda$, where $p_1 > 0$, $\lambda > 0$. We will allow the basic random variables to assume both positive and negative values, but will assume that $p_1 > 0$.

More formally, we are defining a set function for the space $\mathcal{L}_0[0, \infty)$ where $\hat{w} \in \mathcal{L}_0[0, \infty)$ implies that $\hat{w}(t) = w(t) - kt$, $0 \leq t < \infty$, $k > 0$, for $w \in \mathcal{D}_0[0, \infty)$.

Let $\mu_{t_1, t_2, \ldots, t_n}$ be the measure determined by the requirement that

$$P_0[\hat{w}(t_1) \leq x_1, \hat{w}(t_2) \leq x_2, \ldots, \hat{w}(t_n) \leq x_n]$$
$$= P[x(t_1) \leq x_1 + kt_1, x(t_2) \leq x_2 + kt_2, \ldots, x(t_n) \leq x_n + kt_n].$$

Again, this set function satisfies Kolmogorov's symmetry and consistency conditions, and hence determines a unique probability distribution on the Borel sets of $\mathcal{L}_0[0, \infty)$. In [3], Sparre Andersen points out that among the Borel sets of $\mathcal{L}_0[0, \infty)$ are all sets defined by a finite or denumerable number of linear inequalities in the variables $T_1, X_1, T_2, X_2, \ldots$ and hence $\psi(u, T)$ and $\psi(u)$ exist since the condition $k > 0$ guarantees that ruin can only occur at the time of a claim. Thus, $\psi(u, T)$ must be the probability of the event

$$\left\{ u + \sum_{i=1}^{n} (kT_i - X_i) < 0, \sum_{i=1}^{n} T_i \leq T, \quad \text{for at least one value } n = 1, 2, \ldots \right\},$$

and $\psi(u)$ is the probability of the event

$$\left\{ u + \sum_{i=1}^{n} (kT_i - X_i) < 0 \quad \text{for at least one value } n = 1, 2, \ldots \right\}.$$

REFERENCES

1. ABRAMOWITZ, M. and SEGUN, I., Editors (1968). *Handbook of Mathematical Functions.* Dover Publications, New York.
2. AMMETER, H. (1955). The calculation of premium rates for excess of loss and stop loss reinsurance treaties. *Non-proportional Reinsurance,* edited by S. Vajda. Arithbel, S. A., Brussels.
3. ANDERSEN, E. SPARRE (1957). On the collective theory of risk in the case of contagion between the claims. *Transactions XVth International Congress of Actuaries,* New York, Vol. II, 219–229.
4. BARTLETT, D. K. (1965). Excess ratio distribution in risk theory. *Trans. Soc. Actuaries* 17, 435–453.
5. BAXTER, G. and DONSKER, M. (1957). On the distribution of the supremum functional for processes with stationary independent increments. *Trans. Amer. Math. Soc.* 85, 73–87.
6. BEARD, R. E., PENTIKAINEN, T. and PESONEN, E. (1969). *Risk Theory.* Methuen, London.
7. BEEKMAN, J. A. (1966). Research on the collective risk stochastic process. *Skand. Aktuarietidskr.* 49, 65–77.
8. — (1968). Collective risk results. *Trans. Soc. Actuaries* 20, 182–199.
9. — (1969). A ruin function approximation. *Trans. Soc. Actuaries* 21, 41–48.
10. — (1969). Author's review of discussions of 'A Ruin Function Approximation'. *Trans. Society of Actuaries* 21, 277–279.
11. BEEKMAN, J. A. and BOWERS, N. L. (1972). An approximation to the finite time ruin function. *Skand. Aktuarietidskr.* 55, 41–56.
12. — An approximation to the finite time ruin function, Part two, *Skand. Aktuarietidskr.* 55, 128–131.
13. BILLINGSLEY, P. (1968). *Convergence of Probability Measures.* Wiley, New York.
14. BOHMAN, H. and ESSCHER, F. (1963–1964). Studies in risk theory with numerical illustrations concerning distribution functions and stop loss premiums. *Skand. Aktuarietidskr.* 46, 173–225; and 47, 1–40.
15. BOHMAN, H. (1963). To compute the distribution function when the characteristic function is known. *Skand. Aktuarietidskr.* 46, 41–46.
16. — (1971). Ruin probabilities. *Skand. Aktuarietidskr.* 54, 159–163.
17. BOWERS, N. L., JR. (1966). Expansion of probability density functions as a sum of gamma densities with applications in risk theory. *Trans. Soc. Actuaries* 18, 125–137.
18. — (1969). Discussion of 'A Ruin Function Approximation'. *Trans. Society of Actuaries* 21, 275–277.
19. — (1969). An upper bound on the stop-loss net premium. *Trans. Soc. Actuaries* 21, 211–217.

20. BÜHLMANN, H. (1970). *Mathematical Methods in Risk Theory*. Springer-Verlag, Berlin.
21. CRAMÉR, H. (1955). Collective risk theory. *Forsäkringsaktiebolaget Skandia*, Jubilee Volume.
22. FELLER, W. (1968). *An Introduction to Probability Theory and Its Applications*, Vol. 1, 3rd Edition. Wiley, New York.
23. FRYE, W. B. (1974). Collective risk probabilities, random walk, and applications. To appear in *Actuarial Research Clearing House*.
24. GERBER, H. V. (1971). On the discounted compound poisson distribution. To appear in *Proceedings, Wisconsin Actuarial Conference*, 1971.
25. GRANDELL, J. and SEGERDAHL, C.-O. (1971). A comparison of some approximations of ruin probabilities. *Skand. Aktuarietidskr.* **54**, 143–158.
26. HARPER, H. L. (1964). New tables of incomplete gamma-function ratio and of percentage points of the chi-square and beta distributions. U.S. Aerospace Research Laboratories, U.S. Government Printing Office, Washington, D.C.
27. JACKSON, C. J. (1971), Stochastic risk models with investment fluctuations. To appear in *Proceedings, Wisconsin Actuarial Conference*, 1971.
28. KAHN, P. M. (1962). An introduction to collective risk theory and its application to stop-loss reinsurance. *Trans. Soc. Actuaries* **14**, 400–425.
29. MCFALL, M. C. (1972). Net stop-loss premiums. *Actuarial Research Clearing House*, Issue 1972, 3.
30. PEARSON, K. (1965). *Tables of the Incomplete Γ-Function*. University Press, Cambridge.
31. PHILIPSON, C. (1968). A review of the collective theory of risk. *Skand. Aktuarietidskr.* **51**, 1–41.
32. SALVOSA, L. R. (1930). Tables of Pearson's type III function. *Ann. Math. Statist.* **1**, 191–198, plus Appendix 1–187.
33. SEAL, H. (1969). Simulation of the ruin potential of nonlife insurance companies. *Trans. Soc. Actuaries* **21**, 563–585.
34. — (1969). *Stochastic Theory of a Risk Business*. Wiley, New York.
35. SEGERDAHL, C. O. (1959). A survey of results in the collective theory of risk. *Probability and Statistics: The Harald Cramér Volume*, 276–299. Almqvist & Wiksell, Uppsala.
36. SELBY, S. M., Editor (1969). *Standard Mathematical Tables*, 17th Edition. Chemical Rubber Co., Cleveland.
37. SHELLARD, G. D. (1972). Olivetti program of Beekman's approximation. *Actuarial Research Clearing House*, Issue 1972. 1.
38. SPIEGEL, M. R. (1965). *Theory and Problems of Laplace Transforms*. Schaum, New York.
39. TAKÁCS, L. (1965). On the distribution of the supremum of stochastic processes with interchangeable increments. *Trans. Amer. Math. Soc.* **119**, 367–379.
40. THORIN, O. (1971). Analytical steps towards a numerical calculation of the ruin probability for a finite period when the risk process is of the Poisson type or of the more general type studied by Sparre Andersen. *Astin Bulletin* **6**, 54–65.
41. THORIN, O. and WIKSTAD, N. Some numerical values of ruin probabilities

for a finite time period when the claim distribution is of Pareto type. To appear.
42. TITCHMARSH, E. C. (1939). *The Theory of Functions*. Oxford.
43. WIDDER, D. V. (1961). *Advanced Calculus*. Prentice-Hall, Englewood Cliffs, N.J.
44. WIKSTAD, N. (1971). Exemplification of ruin probabilities. *Astin Bulletin* 6, 147–152.
45. WOODDY, J. C. (1963). *Study Note on Risk Theory*. (A monograph published privately by the Society of Actuaries.)

CHAPTER 4

GAUSSIAN MARKOV PROCESSES

4.0. INTRODUCTION

Let us consider a simple simulation of Brownian motion. Use a glass coffee perculator. In addition to placing coffee in the regular unit, drop some of the coffee granules into the water. As the water boils, these granules will simulate the random movements of particles in Brownian motion. Keep this motion in mind as we build a mathematical model for Brownian motion.

To be completely accurate, we would need to build a three dimensional model. However, it is as instructive to focus on one space component of the displacement (from its starting point), say $X(t)$, where $t \geqslant 0$ and represents time. We may imagine that all of the paths are normalized so that $X(0)=0$. Consider a graph of a typical path $X(t)$, $0 \leqslant t \leqslant 1$.

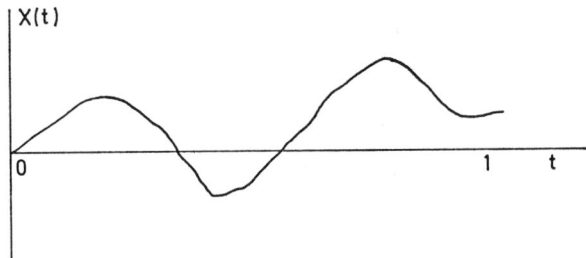

Each random position $X(t_0)$, $0 < t_0 \leqslant 1$, is arrived at by a large number of small movements. Therefore, invoking the Central Limit Theorem, it is natural to say that each $X(t_0)$ has a normal distribution, and that any finite collection of $X(t)$'s have a multivariate normal distribution. It is reasonable, therefore, to call $\{X(t), 0 \leqslant t \leqslant 1\}$ a normal (Gaussian) process. Furthermore, if we know the position of the particle at t_i, our prediction of its position at t_{i+k}, for k a positive integer, will not be affected by any knowledge of the positions at times prior to t_i. This is tantamount to saying that $\{X(t), 0 \leqslant t \leqslant 1\}$ is also a Markov process.

We will now give a more formal presentation. A stochastic process is an infinite collection of random variables $\{x_t, t \in T\}$ which describes the evolution of a natural phenomenon. The process is Gaussian if the joint distribution of every finite set of the x_t's is Gaussian. This means that for any integer n and any subset $\{t_1, t_2, ..., t_n\}$ of T the n random variables $X(t_1), ..., X(t_n)$ have a joint characteristic function given, for any real numbers $u_1, u_2, ..., u_n$, by

$$\varphi_{X(t_1), X(t_2), ..., X(t_n)}(u_1, u_2, ..., u_n)$$
$$= E\{\exp[i(u_1 X(t_1) + ... + u_n X(t_n)]\}$$
$$= \exp\left\{i \sum_{j=1}^{n} u_j E[X(t_j)] - \tfrac{1}{2} \sum_{k=1}^{n} \sum_{j=1}^{n} u_j u_k \operatorname{Cov}[X(t_j), X(t_k)]\right\}.$$

It is clear from the above that the mean value function $E[X(t)]$, $t \in T$, and the covariance kernel $\operatorname{Cov}[X(s), X(t)]$, $s \in T, t \in T$ determine the complete probability law of the process. For $j, k = 1, 2, ..., n$, let $m_j = E(X_j)$, $C_{jk} = \operatorname{Cov}[X_j, X_k]$. Let the matrix of covariances be denoted by C. Then

$$C = \begin{pmatrix} C_{11} & C_{12} & \cdots & C_{1n} \\ C_{21} & C_{22} & \cdots & C_{2n} \\ \vdots & \vdots & \vdots & \vdots \\ C_{n1} & C_{n2} & \cdots & C_{nn} \end{pmatrix}.$$

If C possesses an inverse

$$C^{-1} = \begin{pmatrix} C^{11} & C^{12} & \cdots & C^{1n} \\ C^{21} & C^{22} & \cdots & C^{2n} \\ \vdots & \vdots & \vdots & \vdots \\ C^{n1} & C^{n2} & \cdots & C^{nn} \end{pmatrix},$$

then it can be shown that $X_1, X_2, ..., X_n$ possess a joint probability density given, for any real numbers $x_1, x_2, ..., x_n$ by

$$f_{X_1, X_2, ..., X_n}(x_1, x_2, ...x_n)$$
$$= \frac{1}{(2\pi)^{n/2}} \frac{1}{|C|^{1/2}} \exp\left\{-\tfrac{1}{2} \sum_{k=1}^{n} \sum_{j=1}^{n} (x_j - m_j) C^{jk}(x_k - m_k)\right\},$$

where $|C|$ is the determinant of the matrix C. In some later problems, the student will need to evaluate this density in the Wiener case for $n = 1, 2, 3$.

It is usual to say that a process is Markovian if for any integer $n \geq 1$, if $t_1 < \ldots < t_n$ are parameter values, the conditional x_{t_n} probabilities relative to $x_{t_1}, \ldots, x_{t_{n-1}}$ are the same as those relative to $x_{t_{n-1}}$ in the sense that for each λ

$$P\{x_{t_n}(\omega) \leq \lambda \mid x_{t_1}, \ldots, x_{t_{n-1}}\} = P\{x_{t_n}(\omega) \leq \lambda \mid x_{t_{n-1}}\}$$

with probability 1.

For a Gaussian process to be Markovian the covariance function $r(s, t) = E\{[x(s) - m(s)][x(t) - m(t)]\}$ must be of the factorable form:

(*) $$r(s, t) = \begin{cases} u(s) v(t), & s \leq t \\ u(t) v(s), & t \leq s \end{cases}$$

where $u(t) \geq 0$, $v(t) > 0$, $u(t)/v(t)$ is strictly increasing on $[a, b]$. The conditions on u and v insure that r is a covariance function and that there is a transformation of the process to the Wiener process. Hence almost all sample functions are continuous on $[a, b]$. See Doob [13].

The best known examples of Gaussian Markov processes are the following: (assume that the mean functions are identically zero)

1. Wiener process [37] $u(t) = t$, $v(t) = 1$, $0 \leq t \leq T$
2. Doob-Kac process, See [13], [25]. $u(t) = t$, $v(t) = 1 - t$, $0 \leq t < 1$.
3. Ornstein-Uhlenbeck family of processes. [29]. $u(t) = \sigma^2 e^{\alpha t}$, $v(t) = e^{-\alpha t}$, $\sigma^2 > 0$, $\alpha > 0$, $0 \leq t \leq T$.

The Wiener process has been used to describe the Brownian motion of particles, the movement of the stock market, and in many areas of physics, statistics, etc. The Doob-Kac process is sometimes called the tied-down Wiener process or Brownian bridge because not only does $x(0) = 0$ as in the Wiener process but also $x(1) = 0$. It has been used in computing the distributions of non-parametric statistics, specifically Kolmogorov-Smirnov statistics. The Ornstein-Uhlenbeck process was originally conceived to describe the *velocities* of particles in Brownian motion. Since then, it has been used extensively in electrical engineering and elsewhere.

An interesting way to characterize their sample paths is to say that, with the exceptions of the Wiener process and a few others, a Gaussian Markov process describes the paths of particles subject to *drift*. Thus if we are looking at paths which start at the origin, we could say the process has positive drift if the paths tend to move up into the first quadrant. This can be said more precisely by the use of a partial differential equation.

Let $\{X(t), t \in T\}$ be a Gaussian Markov process with covariance function as (*) and let

$$p(x, s; \alpha \tau) = \frac{\partial}{\partial \alpha} P[x(\tau) \leqslant \alpha \,|\, x(s) = x] = [2\pi A(s, \tau)]^{-\frac{1}{2}} \exp\left\{-\frac{[\alpha - xv(\tau)/v(s)]^2}{2A(s, \tau)}\right\}$$

where $A(s, \tau) = [u(\tau)v(\tau) - u(s)v^2(\tau)/v(s)]$, $a \leqslant s \leqslant \tau \leqslant b$ and the u and v functions are as before. This is called the transition density function of a Gaussian Markov process. Since

$$\lim_{t \to s+} P[x(t) \leqslant y \,|\, x(s) = x] = \begin{cases} 1, y > x \\ 0, y < x, \end{cases}$$

the transition density function determines a process with $x(s) = x$, $x(t) = y$ with probability one. A comparison of $p(x, s; y, t)$ with the usual bivariate conditional normal density reveals that we are assuming an underlying process $\{x(\tau), 0 \leqslant \tau \leqslant T\}$ with zero mean function. The p function satisfies the two diffusion equations of Fokker–Planck which are

$$A(t)\frac{\partial^2 p}{\partial y^2} - B(t)\frac{\partial}{\partial y}[yp] = \frac{\partial p}{\partial t}, \quad 0 \leqslant s < t, y \in R^1$$

$$A(s)\frac{\partial^2 p}{\partial x^2} + xB(s)\frac{\partial p}{\partial x} + \frac{\partial p}{\partial s} = 0, \quad 0 < s < t, x \in R^1$$

where
$$A(t) = [v(t)u'(t) - u(t)v'(t)]/2,$$
$$B(t) = v'(t)/v(t), \quad 0 \leqslant t \leqslant T.$$

The transition function p also fulfills the conditions

(a) $p(x, s; y, t) \geqslant 0$

(b) $\int_{-\infty}^{\infty} p(x, s; y, t)\, dy = 1$

(c) $\lim_{t \to s+} p(x, s; y, t) = 0$ for $y \neq x$.

Note that the "drift" coefficient $yB(t) = 0$ for the Wiener process but not for the Doob–Kac or the Ornstein–Uhlenbeck processes.

We say that a stochastic process has independent increments if for any finite collection of t values, say $t_1 < t_2 < \ldots < t_n$ $(n > 3)$, the differences or increments $Y_{t_2} - Y_{t_1}, \ldots, Y_{t_n} - Y_{t_{n-1}}$ are mutually independent.

A Gaussian Markov process with zero mean function in general does not have independent increments since that would require that for any parameter values t_1, t_2, t_3, t_4 with $t_1 < t_2 < t_3 < t_4$

$$E\{[x_{t_2} - x_{t_1}][x_{t_4} - x_{t_3}]\} = E\{x_{t_2} - x_{t_1}\} E\{x_{t_4} - x_{t_3}\} = 0,$$

whereas all we can conclude is that

$$E\{[x_{t_2} - x_{t_1}][x_{t_4} - x_{t_3}]\} = [v(t_4) - v(t_3)][u(t_2) - u(t_1)],$$

which is non-zero unless $u(t)$ or $v(t)$ is constant, $a \leq t \leq b$. We will see later that a Gaussian Markov process is related to the Wiener process, but it will not usually be by just a change in the time parameter because that assumes that the process has independent increments, See Doob [14], p. 420.

Exercises

1. For the Wiener process, show that $E\{X(s)X(t)\} = $ minimum (s, t).
2. For the Wiener process, verify that the transition density function satisfies the two diffusion equations.

We will now discuss probability measures for Gaussian Markov processes.

Let the basic sample space be $C_0[0, t]$, the set of continuous functions on $[0, t]$ which vanish at the origin. Consider the class $\mathcal{J}$ of quasiintervals (cylinder sets) $I = \{x \in C_0 : a_i < x(t_i) \leq b_i, \ i = 1, 2, \ldots, n\}$ for every choice of integral n and real a_i, b_i. By letting $n = 1, a_1 = -\infty, b_1 = +\infty$, we see that the entire space $C_0[0, t]$ is an I. However, the complement of an I need not be an I. For example, the complement of $I = \{x \in C_0 : -1 < x(\tfrac{1}{2}) \leq 1\}$ is not an I. Furthermore, the union of two I's need not be an I. However, $\mathcal{J}$ can be enlarged to a Borel field of sets, say $\mathcal{J}$.

Recall the definition: A class $\mathcal{J}$ of sets of Ω is a Borel field if

(1) $\Omega \in \mathcal{J}$

(ii) if $\Lambda \in \mathcal{J}$, then $\Omega - \Lambda \in \mathcal{J}$

(iii) if $\Lambda_1, \Lambda_2, \ldots \in \mathcal{J}$, then $\bigcup_{i=1}^{\infty} \Lambda_i \in \mathcal{J}$.

Now consider the set function:

$P[x \in C_0 : a_i < x(t_i) \leq b_i, \ i = 1, 2, \ldots, n]$

$$= \int_{a_n}^{b_n} \cdots \int_{a_2}^{b_2} \int_{a_1}^{b_1} \exp\left\{ -\sum_{i=1}^{n} \frac{[x_i - v(\tau_i) x_{i-1}/v(\tau_{i-1})]^2}{2A(\tau_{i-1}, \tau_i)} \right\}$$

$$\times \left[(2\pi)^n \prod_{i=1}^{n} A(\tau_{i-1}, \tau_i) \right]^{-\frac{1}{2}} dx_1 dx_2 \ldots dx_n \quad \text{where } \tau_0 = 0, x_0 = 0.$$

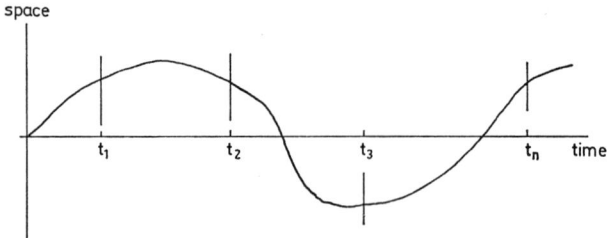

For the Wiener process, $v(t) \equiv 1$, $A(\tau_{i-1}, \tau_i) = \tau_i - \tau_{i-1}$, and the above set function is considerably simpler.

This set function can be used to generate a probability measure $\lambda_{\tau,0}$ on β.

We will use the

DEFINITION. If $\mathcal{J}$ is a Borel field of subsets of Ω, a function μ defined on $\mathcal{J}$ is a probability measure if and only if:

(i) $\infty > \mu(A) \geqslant 0$, $A \in \mathcal{J}$

(ii) $\mu(\Omega) = 1$

(iii) if $\Lambda_i \in \mathcal{J}$, $\Lambda_i \cap \Lambda_j = \emptyset$ for $i \neq j$,

then

$$\mu\left(\bigcup_{i=1}^{\infty} \Lambda_i\right) = \sum_{i=1}^{\infty} \mu(\Lambda_i).$$

A more technically correct procedure would have been to assign the given multivariate normal distribution to the cylinder sets in the class of realvalued functions which vanish at $\tau = 0$. Then observe that this set function satisfies the Kolmogorov consistency conditions:

1. If $\alpha_1, \alpha_2, ..., \alpha_n$ is a permutation of $1, 2, ..., n$ then

$$P[x \in C_0: a_i < x(t_i) \leqslant b_i, \ i = 1, 2, ..., n]$$
$$= P[x \in C_0: a_{\alpha_i} < x(t_{\alpha_i}) \leqslant b_{\alpha_i}, i = 1, 2, ..., n].$$

2. If $m < n$, then

$$P[x \in C_0: -\infty < x(t_i) \leqslant b_i, \quad i = 1, 2, ..., m]$$
$$= \lim_{b_j \to +\infty} P[x \in C_0: -\infty < x(t_i) \leqslant b_i, \quad i = 1, 2, ..., n].$$
$$j = m+1, ..., n$$

Next, adjoin to the cylinder sets the space $C_0[0, t]$ and postulate that $P[x \in C_0[0, t]] = 1$. Construct the Borel field and extend the measure to all sets of the Borel field. This new measure on the set of all real-valued functions is concentrated on $C_0[0, t]$. This procedure is due to J. L. Doob.

This probability measure gives us the mathematical machinery to ask questions such as:

(1) What is $\lambda_{r,0}\{\max_{0 \leq t \leq 1} |x(t)| \leq \alpha\}$ for $\alpha > 0$?
(2) What is $\lambda_{r,0}\{a \leq x(t) \leq b, \ 0 \leq t \leq 1\}$?

Each question has as its answer a real number attached to a set in $\mathcal{J}$.

Exercises

All of the problems refer to the Wiener process.

1. Compute $P[x(\cdot) : x(0) = 0, \ 2 < x(\frac{1}{4}) \leq 3]$.
   We thus see that Wiener measure does not assign significant measure to some sets which appear to be rich in functions.
2. Use the fact that $A \subseteq B$ implies $P(A) \leq P(B)$ to compute
   $$P[x(\cdot): x(0)=0, \ 2<x(\tfrac{1}{4}) \leq 3, \ -3<x(\tfrac{1}{2}) \leq -2, \ 2<x(\tfrac{3}{4}) \leq 3].$$
3. Compute $P[x(\cdot): x(0)=0, \ 0<x(\tfrac{1}{4}) \leq 1]$.
4. Derive Simpson's Rule for Iterated Integrals.

$$\int_b^{b+2k} \int_a^{a+2h} f(x, y)\, dx\, dy$$

$$= \frac{hk}{9} [f(a, b) + 4f(a+h, b) + f(a+2h, b)]$$

$$+ \frac{4hk}{9} [f(a, b+k) + 4f(a+h, b+k) + f(a+2h, b+k)]$$

$$+ \frac{hk}{9} [f(a, b+2k) + 4f(a+h, b+2k) + f(a+2h, b+2k)],$$

provided $f(x, y)$ is a polynomial of degree ≤ 3 in x for each fixed y and a polynomial of degree ≤ 3 in y for each fixed x. Recall the one-dimensional Simpson's Rule. If $f(x)$ is a polynomial of degree ≤ 3,

$$\int_a^{a+2h} f(x)\, dx = \frac{h}{3}[f(a) + 4f(a+h) + f(a+2h)].$$

5. The problem 4 to approximate $P[x(\cdot): x(0)=0, 0<x(\tfrac{1}{4})\leqslant 1, 0<x(\tfrac{1}{2})\leqslant 1]$.
6. Approximate $P[x(\cdot):x(0)=0, \ 0<x(\tfrac{1}{4})\leqslant 1, \ 0<x(\tfrac{1}{2})\leqslant 1, \ -1<x(\tfrac{3}{4})\leqslant 0]$.
   The reader will find it easier to just use the one-dimensional Simpson's rule three times rather than derive a three dimensional version of Problem 4.
7. Compute $P[x(\cdot): x(0)=0, \ -1<x(\tfrac{1}{4})\leqslant 1]$.
8. Approximate

$$P[x(\cdot):x(0)=0, \ -1<x(\tfrac{1}{4})\leqslant 1, \ -1<x(\tfrac{1}{2})\leqslant 1]$$

by applying the two-dimensional Simpson's rule from Problem 4. Motivated by the nonsensical answer, devise an electronic computer program which computes lower and upper Riemann sums for the integral and quits subdividing the square of integration when the upper and lower Riemann sums differ by less than .001.

9. Enlarge your computer program to approximate

$$P[x(\cdot): x(0)=0, \ -1<x(\tfrac{1}{4})\leqslant 1, \ -1<x(\tfrac{1}{2})\leqslant 1, \ -1<x(\tfrac{3}{4})\leqslant 1].$$

For some readers, it is now logical to skip ahead to section 4.7 and then to Chapter 6.

4.1. FUNCTION SPACE INTEGRALS

Much research has been done with function space integrals over the sample paths of Gaussian Markov processes. We will describe them first through a measure approach and then through a limit of sequential integrals.

We need the *Definition*: A (real-valued) functional $F[x]$ on $C_0[0, t]$ assigns a real number to each function $x(\cdot)$ in $C_0[0, t]$. $F[x]$ is a Borel functional if for every Borel set M of the real line, $\{x(\cdot): F[x]\in M\}\in \mathcal{J}$. We speak of such a functional as *measurable* (with respect to the Borel subsets of $C_0[0, t]$).

Example 1. $F[x]=x(\tfrac{1}{2}), \ x\in C_0[0, t]$. If $x_1(\cdot)=\sin(\cdot), \ F[x_1]=\sin(\tfrac{1}{2})$.

Example 2. $G[x]=\int_0^1 x(t)\,dt, \ x\in C_0[0, 1]$. If $x_2(\cdot)=\cos(\cdot), \ G[x_2]=\sin 1$.

Example 3. $H[x]=\sup_{0\leqslant t\leqslant 1}|x(t)|, \ x\in C_0[0, 1]$. If $x_3(t)=t, \ H[x_3]=1$.

Example 4. $I[x]=\sup_{0\leqslant t\leqslant 1} x(t), \ x\in C_0[0, 1]$.

We can now state some results, which are merely special cases of standard results in measure theory textbooks.

Result 1. If the functional $F[x] \geq 0$, and is measurable,

$$I = \int_{C_0[0,t]} F[x] d\mu(x) \text{ exists as an extended real number.}$$

If $I < \infty$, we say $F[x]$ is *integrable* on $C_0[0, t]$.

Rough Definition. The Gaussian Markov integral of a functional $F[x]$ defined on $C_0[0, t]$ is the average value of $F[x]$ computed with respect to the weights that Gaussian Markov measure attaches to the subsets of $C_0[0, t]$. We may use the alternate notation $E_{\lambda_r, 0}\{F[x] | x(0) = 0\}$. If the basic space consisted only of $x_1(\cdot)$, $x_2(\cdot)$, $x_3(\cdot)$ above, each with weight $\frac{1}{3}$, then $E\{F[x]\} = F[x_1]/3 + F[x_2]/3 + F[x_3]/3$.

Result 2. If $F[x]$ is measurable and is expressible as $F[x] = G[x] - H[x]$, $x \in C_0[0, t]$, where G and H are extended positive valued *integrable* functions on $C_0[0, t]$, then $F[x]$ is integrable on $C_0[0, t]$.

Result 3. If $\{F_n[x]\}$ are extended positive valued measurable functionals, then

$$\int_{C_0[0,t]} \sum_{i=1}^{\infty} F_i[x] d\mu(x) = \sum_{i=1}^{\infty} \int_{C_0[0,t]} F_i[x] d\mu(x).$$

Result 4. If

$$F[x] \geq 0, \int_{C_0[0,t]} F[x] d\mu(x) \geq 0.$$

Result 5. If $F[x]$ and $G[x]$ are integrable functionals with $F[x] \geq G[x]$ for almost every

$$x(\cdot) \in C_0[0, t], \int_{C_0[0,t]} F[x] d\mu(x) \geq \int_{C_0[0,t]} G[x] d\mu(x).$$

Result 6. If $F[x]$ is an integrable functional, then

$$\left| \int_{C_0[0,t]} F[x] d\mu(x) \right| \leq \int_{C_0[0,t]} |F[x]| d\mu(x).$$

Result 7. If $F[x]$ and $G[x]$ are integrable functionals, then

$$\int_{C_0[0,t]} |F[x] + G[x]| d\mu(x) \leq \int_{C_0[0,t]} |F[x]| d\mu(x) + \int_{C_0[0,t]} |G[x]| d\mu(x).$$

Result 8. If $F[x]$ is an integrable functional, α and β are real numbers, and E is a measurable subset of $C_0[0, t]$ such that, for $x(\cdot) \in E$, $\alpha \leqslant F[x] \leqslant \beta$, then

$$\alpha \mu(E) \leqslant \int_E F[x] \, d\mu(x) \leqslant \beta \mu(E).$$

Result 9. If $F[x]$ is an a.e. non-negative integrable functional, then a N. and S. condition that

$$\int_{C_0[0,\,t]} F[x] \, d\mu(x) = 0$$

is that $F[x] = 0$ a.e.

Result 10. If $F[x]$ is an integrable functional and E is a set of measure zero, then

$$\int_E F[x] \, d\mu(x) = 0.$$

Result 11. Monotone convergence theorem. If $\{F_n[x]\}$ is an increasing sequence of extended real valued non-negative measureable functions and if $\lim_{n \to \infty} F_n[x] = F[x]$, for almost every $x \in C_0[0, t]$ then

$$\lim_{n \to \infty} \int_{C_0[0,\,t]} F_n[x] \, d\mu(x) = \int_{C_0[0,\,t]} F[x] \, d\mu(x).$$

Result 12. Lebesgue's dominated convergence theorem. Let $\{F_n[x]\}$ be a sequence of measurable functionals such that $\lim_{n \to \infty} F_n[x]$ exists a.e. Assume there exists an integrable functional $G[x]$ such that $|F_n[x]| \leqslant G[x]$ a.e. Then

$$\lim_{n \to \infty} \int_{C_0[0,\,t]} F_n[x] \, d\mu(x) = \int_{C_0[0,\,t]} \lim_{n \to \infty} F_n[x] \, d\mu(x).$$

THEOREM 1. Let $f(u_1, u_2, \ldots, u_n)$ be Lebesgue measurable on Euclidean n-space. Assume that $0 < t_1 < t_2 < \ldots < t_n \leqslant T$. Then $f[x(t_1), x(t_2), \ldots, x(t_n)]$ is Gaussian Markov measurable, and

$$E_{\lambda_{r,0}}\{f[x(t_1), x(t_2), \ldots, x(t_n)] \mid x(0) = 0\}$$

$$= \int_{-\infty}^{\infty} \overset{(n)}{\cdots} \int_{-\infty}^{\infty} f(u_1, u_2, \ldots, u_n) \exp\left\{-\sum_{i=1}^{n} \frac{[u_i - u_{i-1} v(t_i)/v(t_{i-1})]^2}{2A(t_{i-1}, t_i)}\right\}$$

$$\times \left[(2\pi)^n \prod_{i=1}^{n} [A(t_{i-1}, t_i)]\right]^{-\frac{1}{2}} du_1 \, du_2 \ldots du_n \quad \text{where } u_0 = 0, t_0 = 0.$$

Example 1. Assume that $n=1$ and that $f(u_1)=u_1$.

$$E_{\lambda_{r,0}}\{x(t)|x(0)=0\} = \int_{-\infty}^{\infty} \frac{u\,e^{-u^2/[2A(0,t)]}}{\sqrt{2\pi A(0,t)}}\,du = 0$$

Example 2

$$\text{Var}\,\{x(t)\} = E_{\lambda_{r,0}}[x(t)^2] = \int_{-\infty}^{\infty} \frac{u^2 e^{-u^2/[2A(0,t)]}}{\sqrt{2\pi A(0,t)}}\,du = A(0,t)$$

Example 3. If $s<t$,

$$E\{x(s)\,x(t)|x(0)=0\} = \text{Cov}\,((x(s),x(t)))$$

$$\int_{-\infty}^{\infty}\int_{-\infty}^{\infty} x_s x_t \exp\left\{-\frac{x_s^2}{2A(o,s)} - \frac{[x_t - x_s v(t)/v(s)]^2}{2A(s,t)}\right\}$$

$$\times [(2\pi)^2 A(o,s)\,A(s,t)]^{-\frac{1}{2}}\,dx_s\,dx_t = u(s)\,v(t).$$

For the Wiener process, this can be stated: $\text{Cov}\,\{x(s),x(t)\} = \min(s,t)$.

Exercises

1. Compute the following Wiener integrals:

   a. $\int_{C_0[0,0]} x(\tfrac{1}{2})\,d_w x$

   b. $\int_{C_0[0,1]} x^2(\tfrac{3}{4})\,d_w x$

   c. $\int_{C_0[0,1]} x(\tfrac{1}{2})\,x(\tfrac{3}{4})\,d_w x$

   d. $\int_{C_0[0,1]} \{x(\tfrac{1}{4}) + x^2(\tfrac{1}{2}) + x(\tfrac{3}{4})\,x(\tfrac{7}{8})\}\,d_w x$

2. Prove the following

   THEOREM. Let $0 \leqslant t_1 < t_2 < t_3 < t_4 \leqslant 1$. Then

   $$\int_{C_0[0,1]} f[x(t_2)-x(t_1)]\,g[x(t_4)-x(t_3)]\,d_w x$$

   $$e \int_{C_0[0,1]} f[x(t_2)-x(t_1)]\,d_w x \int_{C_0[0,1]} g[x(t_4)-x(t_3)]\,d_w x.$$

The equality e means if one side exists (as a finite number), they both do and are equal.

3. Use the above theorem to compute

   a. $\int_{C_0[0,1]} \{x(\tfrac{1}{2}) - x(\tfrac{1}{4})\}\{x(1) - x(\tfrac{3}{4})\} \, d_w x$

   b. $\int_{C_0[0,1]} \{x(\tfrac{1}{2}) - x(\tfrac{1}{4})\}^2 \{x(1) - x(\tfrac{3}{4})\}^2 \, d_w x$

4. Compute the following

   a. $\int_{C_0[0,1]} x(\tfrac{1}{4}) \, x(\tfrac{1}{2}) \, x(\tfrac{3}{4}) \, d_w x$

   b. $\int_{C_0[0,1]} x(\tfrac{1}{4}) \, x(\tfrac{1}{2}) \, x(\tfrac{3}{4}) \, x(1) \, d_w x$

   c. $\int_{C_0[0,1]} x(\tfrac{1}{5}) \, x(\tfrac{2}{5}) \, x(\tfrac{3}{5}) \, x(\tfrac{4}{5}) \, x(1) \, d_w x$

There are some Gaussian–Markov integrals which can easily be calculated by interchanging the order of integration. This technique depends on a deep theorem known as Fubini's Theorem. To explain this theorem, we will need some preliminary definitions.

A Metric for $C_0[0, t]$. Let $N_U(f) = \max_{0 \leq s \leq t} |f(s)|$. This norm yields a metric for $C_0[0, t]$, $d_U(f, g) = \max_{0 \leq s \leq t} |f(s) - g(s)|$, and d satisfies the usual requirements:

$$0 \leq d(f, g) = d(g, f)$$

$$d(f, g) = 0 \quad \text{iff } f = g$$

$$d(f, h) \leq d(f, g) + d(g, h).$$

DEFINITION. A functional $F[x]$ defined on $C_0[0, t]$ is continuous (in the uniform topology) if for any $g \in C_0[0, t]$ and $\varepsilon > 0$, $\exists \delta > 0 \ni d(g, f) < \delta \Rightarrow |F[g] - F[f]| < \varepsilon$.

LEMMA 2. *If* $F[x]$ *is continuous in the uniform topology, it is measurable.*

LEMMA 3. *Let* $F[g; p]$ *be a functional defined for* $g \in C_0[0, t]$, $0 \leq p \leq t$. *Assume that it is continuous on the product space* $C_0[0, t] \times [0, t]$. *Then* $F[g; p]$ *is product measurable over* $C_0[0, t] \times [0, t]$.

Remark. Since $x(p)$ is continuous on $C_0[0, t] \times [0, t]$, it is product measurable.

Thus the functional $x^k(t)$ is Gaussian Markov X Lebesgue measurable, for k a positive integer.

Fubini theorem. If $F[g; p]$ is an integrable function on $C_0[0, t] \times [0, t]$, then

$$\int_{C_0[0, t] \times [0,t]} F[g; p] d\mu(x) \, dp$$
$$= \int_{C_0[0, t]} \int_0^t F[g; p] d\mu(x) \, dp$$
$$= \int_0^t \int_{C_0[0, t]} F[g; p] \, dp \, d\mu(x).$$

Example 4.

$$\int_{C_0[0, 1]} \int_0^1 x(t) \, dt \, d_{\lambda_r} x = \int_0^1 \int_{C_0[0, 1]} x(t) \, d_{\lambda_r} x \, dt$$

by the Fubini Theorem

$$= \int_0^1 0 \, dt \quad \text{by our previous example}$$

$$= 0.$$

Example 5.

$$\int_{C_0[0, 1]} \int_0^1 x^2(t) \, dt \, d_{\lambda_r} x = \int_0^1 A(0, t) \, dt$$

by the same technique.

For the Wiener process, $A(0, t) = t$ and the above $= \frac{1}{2}$.

Example 6.

$$\int_{C_0[0, 1]} \left[\int_0^1 x(t) \, dt\right]^2 d_{\lambda_r} x = \int_{C_0[0, 1]} \int_0^1 x(t) \, dt \int_0^1 x(s) \, ds \, d_{\lambda_r} x$$
$$= \int_0^1 \int_0^1 \int_{C_0[0, 1]} x(t) \, x(s) \, d_{\lambda_r} x \, dt \, ds$$
$$= \int_0^1 \int_0^1 r(s, t) \, ds \, dt$$

For the Wiener process $r(s, t) = \min(s, t)$, and the integral equals $\frac{1}{3}$.

Example 7. Consider
$$\int_{C_0[0,\,t]} \exp\left\{\frac{1}{z}\int_0^t p(s)\,x^2(s)\,ds\right\} d\lambda_r\, x$$

where $p(s)$ is positive and continuous, $0 \leq s \leq t$. We will use the Kac–Siegert representation of a Gaussian process (See [22]). By this representation,

$$x(s) = \sum_{k=1}^{\infty} \frac{\alpha_k w_k(s)}{\sqrt{p(s)\,\theta_k}}, \quad 0 \leq s \leq t \quad \text{where} \quad \alpha_1, \alpha_2, \ldots$$

are independent, Gaussian random variables with means 0, variances 1, and $\{\theta_k\}$ and $\{w_k(s)\}$ are the (positive) eigenvalues and normalized eigenfunctions respectively of the Volterra integral equation

$$(1) \quad w(s) = \theta \int_0^t \sqrt{p(s)}\,\sqrt{p(\tau)}\,r(\tau, s)\,w(\tau)\,d\tau.$$

We allow z to be complex but ask that either z be pure imaginary or

$$\text{Re } z > 2u(t)\,G^2 Pt/v(t) \quad \text{where} \quad P = \max_{0 \leq s \leq t} p(s)$$

and
$$v(s) \leq G, \quad 0 \leq s \leq t.$$

Using Parseval's Theorem, monotone convergence, and the independence of the random variables,

$$\int_{C_0[0,\,t]} \exp\left\{\frac{1}{z}\int_0^t p(s)\,x^2(s)\,ds\right\} d\lambda_{r,0}\,x$$

$$= E_{\lambda_{r,0}}\left\{\exp\left[\frac{1}{z}\sum_{k=1}^{\infty}\frac{\alpha_k^2}{\theta_k}\right]\right\}$$

$$= \lim_{N \to \infty} E_{\lambda_{r,0}}\left\{\exp\left[\frac{1}{z}\sum_{k=1}^{N}\frac{\alpha_k^2}{\theta_k}\right]\right\}$$

$$= \lim_{N \to \infty} \prod_{k=1}^{N} \frac{1}{\sqrt{2\pi}} \int_{-\infty}^{\infty} \exp\left[-\frac{y^2}{2}\left(1 - \frac{2}{z\theta_k}\right)\right] dy$$

$$= \lim_{N \to \infty} \prod_{k=1}^{N} \left[1 - \frac{2}{z\theta_k}\right]^{-\frac{1}{2}}$$

$$= [D_r(2/z)]^{-\frac{1}{2}}$$

where $D_r(\theta)$ is the Fredholm determinant associated with equation (1).

Now consider the Wiener process with $p(s) \equiv 1$, $0 \leq s \leq t$. The Fredholm determinant can be determined from an initial value equation

$$\varphi''(t) + \theta \varphi(t) = 0$$
$$\varphi(0) = 0$$
$$\varphi'(0) = 1$$

which has a unique solution $\varphi_\theta(t)$.

$$D_w(\theta) = \frac{\varphi_\theta(1)}{\varphi_0(1)} = \prod_{i=1}^{\infty}\left[1 - \frac{\theta}{\theta_i}\right]$$

where the θ_i are the eigenvalues associated with $r(s, t) = \min(s, t)$. Equation (1) now becomes

$$w(s) = \theta \int_0^t \min(\tau, s) w(\tau) d\tau$$

$$w(s) = \theta \int_0^s \tau w(\tau) d\tau + \theta s \int_s^t w(\tau) d\tau.$$

Hence $w(0) = 0$, $w'(s) = \theta \int_s^t w(\tau) d\tau$.
Hence $w'(t) = 0$.
Furthermore $w''(s) = -w(s)\theta$.

The general solution of this is $w(s) = A \sin \sqrt{\theta} s + B \cos \sqrt{\theta} s$. But $w(0) = 0$ drops $B \cos \sqrt{\theta} s$. The fact that $w'(t) = 0$ implies that

$$\theta_k = (k + \tfrac{1}{2})^2 \frac{\pi^2}{t^2}, \quad k = 0, 1, 2, \ldots$$

The normalization yields

$$w_k(s) = \sqrt{\frac{2}{t}} \sin\left[(k + \tfrac{1}{2})\frac{\pi}{t}\right]s, \quad k = 0, 1, 2, \ldots$$

From an infinite product result

$$D_w\left(\frac{2}{z}\right) = \cosh\left(i\sqrt{\frac{2}{z}}\right)t.$$

If z is not pure imaginary, z need only be such that $\text{Re } z > 2$. Furthermore if z is real,

$$\left[D_w\left(\frac{2}{z}\right)\right]^{-\frac{1}{2}} = \left[\sec\left(\sqrt{\frac{2}{z}}t\right)\right]^{\frac{1}{2}}.$$

Probably the most useful interpretation comes when z is pure imaginary, in which case the function space integral is the characteristic function of the random variable

$$\int_0^t x^2(s)\,ds.$$

If $1/z = i\mu$, we obtain by inversion

$$\lambda_w\left\{\int_0^t x^2(s)\,ds < \beta\right\} = \frac{1}{2\pi}\int_{-\infty}^\beta \int_{-\infty}^\infty e^{-ix\mu}[\operatorname{sech}(i\sqrt{2\mu i})\,t]^{\frac{1}{2}}d\mu\,dx.$$

This expresses the measure of an L_2 neighborhood of the origin.

We will work out the next example using a powerful theorem which has come to be called the "Cameron–Martin Translation Theorem for the Wiener Process". See [6]. Let $\{C, B, \lambda_w\}$ be the Wiener process on $[0, 1]$ with

$$r(s, t) = \min(s, t) = \begin{cases} s, & s \leqslant t \\ t, & s \geqslant t \end{cases}.$$

Let x_0 be any function having a derivative of bounded variation on $[0, 1]$ and such that $x_0(0) = 0$. Then for all (B) measurable functions F,

$$E_{\lambda_w}\{F[x]\} = E_{\lambda_w}\left\{F[x + x_0]\exp\left[-\int_0^1 x_0'(t)\,dx(t) - (\tfrac{1}{2})\int_0^1 [x_0'(t)]^2\,dt\right]\right\}.$$

As an example of this theorem let $F[x] \equiv 1$, $x_0(t) = (\tfrac{1}{2})(1-t)^2 - \tfrac{1}{2}$. We then easily see that

$$E\left\{\exp\left[\int_0^1 x(t)\,dt\right]\,\Big|\,x(0) = 0\right\} = e^{-1/6} \quad \text{since } E\{1\} = 1.$$

Dale Varberg has extended this theorem to Gaussian processes and has given a special translation theorem for Gaussian Markov processes in [35]. This can be stated as follows. Let $\{C, B, \lambda_{rm}\}$ be a Gaussian Markov process on $[a, b]$ with r of the factorable form, and C the space of continuous functions on $[a, b]$. Let x_0 be any function having a derivative of bounded variation on $[a, b]$ and such that $x_0(a) = 0$ if $u(a) = 0$. Then for all (B) measurable functions F, $E\{F[x]\,|\,x(0) = 0\} = E\{F[x + x_0]\,J[x]\,|\,x(0) = 0\}$ where

$$J[x] = \exp\Big\{D_1 + D_2[x(a) - m(a)]$$
$$-\tfrac{1}{2}\int_a^b \frac{v(t)x_0^1(t) - x_0(t)v'(t)}{v(t)u'(t) - u(t)v'(t)}\,d\left[\frac{2x(t) - 2m(t) + x_0(t)}{v(t)}\right]\Big\}$$

and

$$D_1 = \{-x_0^2(a)/[2u(a)\,v(a)]\} \quad \text{if } u(a) \neq 0, \quad \text{and } D_1 = 0 \quad \text{if } u(a) = 0,$$
$$D_2 = \{-x_0(a)/v(a)\} \quad \text{if } u(a) \neq 0, \text{ and } D_2 = 0 \text{ if } u(a) = 0.$$

Let us apply this in the case that the stochastic process is the Ornstein-Uhlenbeck process on $[0, 1]$, the mean function $m(t) \equiv 0, 0 \leq t \leq 1$, $x(0) = 0$, $F[x] \equiv 1$, and $x_0(t) = (\tfrac{1}{2})(1-t)^2$.
Then

$$J[x] = \exp\left\{-1/(4\sigma^2) - \int_0^1 \frac{e^{-\alpha t}(1-t)[1+(\tfrac{1}{2})(1-t)\,\alpha]}{2\sigma^2 \alpha}\right.$$
$$\left. \times d\left[\frac{2x(t) + (\tfrac{1}{2})(1-t)^2}{\exp(-\alpha t)}\right]\right\}$$
$$= \exp\{-1/(4\sigma^2) + (1+\alpha/2)/[4\sigma^2 \alpha]\}$$
$$= \frac{1}{2\sigma^2 \alpha} \int_0^1 [2x(t) + \tfrac{1}{2}(1-t)^2]\left\{\frac{\alpha^2}{2}(1-t)^2 + 2\alpha(1-t) + 1\right\} dt.$$

Thus

$$E\left\{\exp\left\{-\frac{1}{\sigma^2 \alpha}\int_0^1 x(t)\left[\frac{\alpha^2}{2}(1-t)^2 + 2\alpha(1-t) + 1\right] dt\right\} \Big| x(0) = 0\right\}$$
$$= \exp\left\{\frac{1}{2\sigma^2}\left[\frac{\alpha}{20} + \frac{1}{2} - \frac{1}{3\alpha}\right]\right\}.$$

4.2. Sequential Integrals

There is an alternate way of looking at integrals defined over the sample functions of a stochastic process. This method has considerable intuitive appeal. Furthermore, the results agree with the integrals based on measures.

Recall that the transition density function $p(x, s; y, t)$ determines a process with $x(s) = x$, $x(t) = y$ with probability one. Denote the expected value of a functional $F[X]$ for this process by $E\{F[X] | X(s) = x, X(t) = y\}$. Expectations not tied down at t can be obtained through the equation

$$E\{F[X] | X(s) = x\} = \int_{-\infty}^{\infty} E\{F[X] | X(s) = x, X(t) = y\}\, p(x, s; y, t)\, dy.$$

For later purposes, let $p^*(x, s; y, t)$ equal $p(x, s; y, t)$ with $A(s, t)$ replaced by $A(s, t)/\lambda$ where $\text{Re } \lambda \geq 0$, $\lambda \neq 0$.

Let $\tau = [\tau_1, \ldots, \tau_n]$ be a variable vector of a variable number of dimensions whose components form a subdivision of $[s, t]$ so that

$$\tau_0 \equiv s < \tau_1 < \tau_2 < \ldots < \tau_n \equiv t.$$

Let $$\|\tau\| = \max_{j=1,2,\ldots,n}(\tau_j - \tau_{j-1}).$$

Let $\xi \equiv [\xi_1, \ldots, \xi_{n-1}]$ denote an unrestricted real vector, where n is determined by τ, and let $\xi_0 \equiv x$, $\xi_n \equiv y$. Let $\psi_{\tau,\xi}(\tau_i) = \xi_i$, $i = 0, 1, \ldots, n$ and $\psi_{\tau,\xi}$ be linear on $[\tau_{i-1}, \tau_i]$.

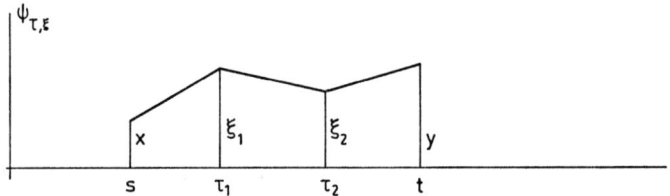

Then we define the sequential Gaussian Markov integral

$$E_\lambda^s\{F[X] \mid X(s) = x, X(t) = y\} = \lim_{\|\tau\| \to 0} \int_{R_{n-1}} G_\lambda(\tau, \xi) F(\psi_{\tau,\xi}) d\xi$$

where $$G_\lambda(\tau, \xi) = \frac{1}{p^*(x, s; y, t)} \prod_{i=1}^n p^*(\xi_{i-1}, \tau_{i-1}; \xi_i, \tau_i).$$

Assume that ξ_n is unrestricted. Then we make the definition

$$E_\lambda^s\{F[X] \mid X(s) = x\} = \lim_{\|\tau\| \to 0} \int_{R_n} \prod_{i=1}^n p^*(\xi_{i-1}, \tau_{i-1}; \xi_i, \tau_i) F(\psi_{\lambda,\xi}) d\xi.$$

Clearly

$$E_\lambda^s\{F[X] \mid X(s) = x\} = \int_{-\infty}^{\infty} E_\lambda^s\{F[X] \mid X(s) = x, X(t) = y\} p^*(x, s; y, t) dy.$$

Let $C[x, s; y, t]$ denote the space of continuous functions with x and y endpoints. For $X \in C[x, s; y, t]$, let $\|X\| = \sup_{s \leq \tau \leq t} |X(\tau)|$. Let $C_p[x, s; y, t]$ denote the set of all polygonal functions in $C[x, s; y, t]$.

A subset S of $C[x, s; y, t]$ is a Borel set if it is a member of the smallest σ-field containing the quasi-intervals

$$\{X \in C[x, s; y, t] : \alpha_i < X(\tau_i) < \beta_i, \quad i = 1, 2, \ldots, n\}$$

where τ ranges over all subdivision vectors of $[s, t]$ and α_i, β_i range over the extended reals. $F[X]$ is a Borel functional if it is measurable with respect to the σ-field of Borel measurable subsets of $C[x, s; y, t]$.

Reference [3] contains the following theorem:

THEOREM 2. Let $F[X]$ be Borel measurable over $C[x, s; y, t]$ and continuous in the uniform topology almost everywhere (in the Gaussian sense) on the space $C[x, s; y, t]$.

Let $\Phi(w)$ be a positive monotone increasing function such that

$$\Phi\left(G\sqrt{\frac{u(t)}{v(t)} - \frac{u(s)}{v(s)}}\, w + |x|\frac{G}{g}\right) \exp[-w^2/2]$$

is integrable on $[0, \infty)$.

Then if $|F(x)| \leq \Phi(\|X\|)$ on $C_p[x, s; y, t]$, the sequential Gaussian Markov integral of F exists for $\lambda = 1$ and equals the Gaussian Markov integral $E_1^s\{F[X]\,|\,X(s)=x, X(t)=y\} = E\{F[X]\,|\,X(s)=x, X(t)=y\}$.

To illustrate sequential integrals, we will consider the following example: For $s \leq \theta \leq t$,

$$E\{X(\theta)|\,X(s)=x, X(t)=y\} = x\frac{v(\theta)}{v(s)}\frac{A(\theta, t)}{A(s, t)} + y\frac{v(t)}{v(\theta)}\frac{A(s, \theta)}{A(s, t)}.$$

Proof. By the Theorem

$$E\{X(\theta)|\,X(s)=x, X(t)=y\} = E_1^s\{X(\theta)|\,X(s)=x, X(t)=y\}$$

$$= \lim_{\|\tau\|\to 0}\int_{R_{n-1}} G(\tau, \xi)\,\psi_{\tau,\xi}\,d\xi$$

where $\psi_{\tau,\xi}(\theta) = \xi_{i-1} + (\theta - \tau_{i-1})\left(\frac{\xi_i - \xi_{i-1}}{\tau_i - \tau_{i-1}}\right),\ \tau_{i-1} \leq \theta \leq \tau_i$.

By repeated application of the Chapman-Kolmogorov equation, both before and after i,

$$\int_{R_{n-1}} G(\tau, \xi)\,\psi_{\tau,\xi}\,d\xi$$

$$= \int_{-\infty}^{\infty}\int_{-\infty}^{\infty} \frac{p(x, s; \xi_{i-1}, \tau_{i-1})\,p(\xi_{i-1}, \tau_{i-1}; \xi_i, \tau_i)\,p(\xi_i, \tau_i; y, t)}{p(x, s; y, t)}$$

$$\times\left[\xi_{i-1}\left(\frac{\tau_i - \theta}{\tau_i - \tau_{i-1}}\right) + \xi_i\left(\frac{\theta - \tau_{i-1}}{\tau_i - \tau_{i-1}}\right)\right]d\xi_{i-1}\,d\xi_i.$$

Now consider
$$I = \int_{-\infty}^{\infty} \xi_k p(\xi_{k-1}, \tau_{k-1}; \xi_k, \tau_k) p(\xi_k, \tau_k; \xi_{k+1}, \tau_{k+1}) d\xi_k.$$

By completing the square, one can show
$$I = \frac{\beta}{\alpha} p(\xi_{k-1}, \tau_{k-1}; \xi_{k+1}, \tau_{k+1})$$

where
$$\beta = \frac{v(\tau_k)}{v(\tau_{k-1})} \xi_{k-1} A(\tau_k, \tau_{k+1}) + \frac{v(\tau_{k+1})}{v(\tau_k)} \xi_{k+1} A(\tau_{k-1}, \tau_k)$$

and
$$\alpha = A(\tau_{k-1}, \tau_{k+1}).$$

Using this fact, one obtains
$$\int_{R^{n-1}} G(\tau, \xi) \psi_{\tau, \xi} d\xi = \frac{\tau_i - \theta}{\tau_i - \tau_{i-1}} h(\tau_{i-1}) + \frac{\theta - \tau_{i-1}}{\tau_i - \tau_{i-1}} h(\tau_i)$$

where
$$h(\tau_k) = x \frac{v(\tau_k)}{v(s)} \frac{A(\tau_k, \tau_{k+1})}{A(s, \tau_{k+1})} + x \frac{v(\tau_{k+1})}{v(\tau_k)} \frac{A(s, \tau_k)}{A(s, \tau_{k+1})} \frac{v(\tau_{k+1})}{v(s)} \frac{A(\tau_{k+1}, t)}{A(s, t)}$$
$$+ y \frac{v(t)}{v(\tau_k)} \frac{A(s, \tau_k)}{A(s, t)}.$$

Since the u and v functions are continuous and v is bounded away from 0, for $\varepsilon > 0$, there exists $\delta > 0$ such that $\|\tau\| < \delta$ implies that
$$i(\theta) - \varepsilon < h(\tau_k) < i(\theta) + \varepsilon, \; k = i-1, i$$

where
$$i(\theta) = x \frac{v(\theta)}{v(s)} \frac{A(\theta, t)}{A(s, t)} + y \frac{v(t)}{v(\theta)} \frac{A(s, \theta)}{A(s, t)}.$$

Hence, for that δ,
$$\frac{\tau_i - \theta}{\tau_i - \tau_{i-1}} [i(\theta) - \varepsilon] + \frac{\theta - \tau_{i-1}}{\tau_i - \tau_{i-1}} [i(\theta) - \varepsilon]$$
$$< \int_{R^{n-1}} G(\tau, \xi) \psi_{\tau, \xi} d\xi$$

$$< \frac{\tau_i - \theta}{\tau_i - \tau_{i-1}} [i(\theta) + \varepsilon] + \frac{\theta - \tau_{i-1}}{\tau_i - \tau_{i-1}} [i(\theta) + \varepsilon],$$

or
$$i(\theta) - \varepsilon < \int_{R_{n-1}} G(\tau, \xi)\, \psi_{\tau, \xi}\, d\xi < i(\theta) + \varepsilon.$$

Letting $\varepsilon \to 0$ gives the desired result.

Rather than considering a general process, we will consider the Wiener process and give the following theorem:

THEOREM 3. The Wiener process, conditioned by $X(s) = x$, $X(t) = y$, is Gaussian Markov with mean function

$$m(\theta) = E\{X(\theta) \mid X(s) = x, X(t) = y\} = x + \frac{\theta - s}{t - s}(y - x)$$

and covariance function

$$E\{[X(a) - m(a)][X(b) - m(b)] \mid X(s) = x, X(t) = y\}$$

$$= \begin{cases} \dfrac{(a-s)(t-b)}{t-s}, & a < b \\ \dfrac{(b-s)(t-a)}{t-s}, & b < a \end{cases}$$

A large body of literature has grown up for Gaussian Markov integrals which are not conditioned at the right hand time extremity. This is not so for integrals conditioned at the right. The following theorem is designed to remedy this situation. See [4]. We record some needed definitions.

Let $C[s, t]$ denote the space of continuous functions with $X(s) = 0$. For $X \in C[x, s; y, t]$ or $X \in C[s, t]$, let $\|X\| = \sup_{s \leqslant \tau \leqslant t} |X(\tau)|$.

THEOREM 4. Let Λ be any open set of complex numbers $\lambda \in \operatorname{Re} \lambda > 0$, $\lambda \neq 0$. Let Λ^* denote the closure of Λ with $\lambda = 0$ omitted. Let $p(x, s; y, t)$, $p^*(x, s; y, t)$ and their related integrals be as specified earlier. Let $F[X]$ be a Borel functional for $X \in C[x, s; y, t]$. Assume F also satisfies the following two conditions:

(1) $F[X]$ is a continuous function of X in the uniform topology almost everywhere in $C[x, s; y, t]$.
(2) For all X in $C[x, s; y, t]$, $|F[X]| \leqslant A \exp(M\|X\|^\gamma)$ where A and M are given positive integers, and $0 < \gamma < 2$.

Then for $\lambda \in \Lambda^*$, and functionals for which the right side exists and is analytic in Λ, and continuous in Λ^*,

$$E^s_\lambda\{F[X]\,|\,X(s)=x,\,X(t)=y\}\,p^*(x,s;y,t)$$

$$=\frac{1}{2\pi}\int_{-\infty}^{\infty}\exp\left\{-i\mu\left[y-x\frac{v(t)}{v(s)}\right]\right\}E\left\{F\left[\lambda^{-\frac{1}{2}}X(\cdot)+x\frac{v(\cdot)}{v(s)}\right]\right.$$

$$\left.\times\exp\left[i\mu X(t)/\sqrt{\lambda}\right]\,|\,X(s)=0\right\}d\mu.$$

This theorem was used to calculate the conditioned integral when

$$F[X]=\exp\left\{-i\int_s^t[X^2(\tau)-f(\tau)\,X(\tau)]\,d\tau\right\}.$$

The expectation within the right hand integral was calculated using the Kac–Siegert representation. See [22]. The final result in the Wiener case and for $\lambda=-i$ was

$$E^s_{-i}\{F[X]\,|\,X(s)=x,\,X(t)=y\}\,p^*(x,s;y,t)$$

$$=\frac{2^{\frac{1}{4}}}{[2\pi i\sin\sqrt{2}\,(t-s)]^{\frac{1}{2}}}$$

$$\exp\left\{\frac{i}{\sqrt{2}\sin\sqrt{2}\,(t-s)}\left\{(x^2+y^2)\cos\sqrt{2}\,(t-s)-2xy\right.\right.$$

$$+\sqrt{2}\,x\int_s^t f(b)\sin\sqrt{2}\,(t-b)\,db+\sqrt{2}\,y\int_s^t f(a)\sin\sqrt{2}\,(a-s)\,da$$

$$-\tfrac{1}{2}\int_s^t\int_s^b \sin\sqrt{2}\,(a-s)\sin\sqrt{2}\,(t-b)\,f(a)\,f(b)\,da\,db$$

$$\left.\left.-\tfrac{1}{2}\int_s^t\int_b^t \sin\sqrt{2}\,(b-s)\sin\sqrt{2}\,(t-a)\,f(a)\,f(b)\,da\,db\right\}\right\}.$$

4.3. Gaussian Markov processes, partial differential equations, and integral equations

Perhaps the appropriate way to begin this section is to quote the opening sentence of an excellent paper on this subject by Mark Kac: "The connections between probability theory on the one hand and differential and integral equations on the other, are so numerous and diverse that the task of presenting them in a comprehensive and connected manner appears almost impossible." (See [26]). Nevertheless, we will discuss some of the interesting papers written on this subject.

An early paper which stimulated considerable work was by R. Feynman, [17], which used a certain function space integral to give a solution to the Schroedinger equation. A later chapter will be devoted to this subject. Partially motivated by the Feynman paper, M. Kac published a paper in 1949 related to calculating Wiener integrals through partial differential equations, [24]. His principal result may be described as follows. Let $\{x(t), x(0)=0, 0 \leqslant t < \infty\}$, be elements of the Wiener space and let $V(x)$ be a piecewise continuous, non-negative function defined for $-\infty < x < \infty$.

Let
$$\sigma(\alpha; t) = \text{Prob}\left\{\int_0^t V[x(\tau)]\,d\tau < \alpha\right\}.$$

Then
$$\int_0^\infty \int_0^\infty \exp(-u\alpha - st)\,d_\alpha\,\sigma(\alpha; t)\,dt = \int_{-\infty}^\infty \psi(x)\,dx,$$

where $\psi(x)$ is the fundamental solution (Green's function) of the differential equation

$$\frac{1}{2}\frac{d^2\psi}{dx^2} - (s + uV(x))\,\psi = 0, \quad x \neq 0$$

subject to the conditions

$$\psi(x) \to 0, \quad x \to \pm\infty$$

$$|\psi'(x)| < M, \quad x \neq 0$$

$$\psi'(0+) - \psi'(0-) = -2$$

One of the examples considered is $V(x) = (1 + \text{signum } x)/2$.

Since signum $x = \begin{cases} 1, & x > 0 \\ -1, & x < 0 \\ 0, & x = 0 \end{cases}$

$$\text{Prob}\left\{\int_0^t V[x(\tau)]\,d\tau < \alpha\right\}$$

is the measure of those x's for which the Lebesgue measure of those τ in $[0, t]$ for which $X(\tau) > 0$ is less that α. This uses the fact that the Lebesgue measure of those τ in $[0, t]$ for which $x(\tau)=0$ is 0. In this case the inversion of the double Laplace transform gives

$$\sigma(\alpha; t) = \frac{2}{\pi} \arcsin\left(\frac{\alpha}{t}\right)^{\frac{1}{2}}, \quad 0 \leqslant \alpha \leqslant t.$$

Another interesting paper was published in 1954 by R. H. Cameron. One of its results was that the Wiener integral

$$G(t, \xi) = \int_{C[0,1]} \exp\left\{\frac{t}{a} \int_0^1 \theta\left[t(1-s), 2\sqrt{\frac{t}{a}}\, x(s) + \xi\right] ds\right\} \sigma[2\sqrt{t/a}\, x(1) + \xi] d_w x$$

solved the boundary value problem

$$\frac{\partial^2 G}{\partial \xi^2} - a \frac{\partial G}{\partial t} + \theta(t, \xi) G = 0 \text{ in } R, \quad \lim_{t \to 0+} G(t, \xi) = \sigma(\xi)$$

for almost all real ξ where "a" is a positive number, R is the strip $R: 0 < t < t_0, -\infty < \xi < \infty$, t_0 is a positive number or $+\infty$, and θ and σ are subject to certain conditions. See [8].

This served as the foundation in Cameron's later work on Feynman integrals, which will be described later.

A backwards time version of this for Gaussian Markov processes was considered in [1].

THEOREM 5. Assume that

$$R(a, b) = \begin{cases} U(a) v(b), & a \leqslant b \\ U(b) v(a), & b \leqslant a \end{cases}$$

where $U(p) = u(p) - u(s)v(p)/v(s)$, $s \leqslant p \leqslant t$, and the u and v functions are as described before. Assume that the complex valued functions $\psi(y, p)$ and ψ_y are continuous and bounded for (y, p) such that y is in R_1, $0 \leqslant p \leqslant t$. Assume that $\sigma(y)$ is a complex valued, integrable function of y, with a continuous, bounded first derivative for y in R_1.

Then the Gaussian Markov expectation

$$H(\xi, s) = \int_{C[s, t]} \exp\left\{\int_s^t \psi[x(p) + \xi v(p)/v(s), p] dp\right\} \sigma[x(t) + \xi v(t)/v(s)] d_R x$$

satisfies the partial differential equation

$$\frac{u'(s) v(s) - u(s) v'(s)}{2} \frac{\partial^2 H}{\partial \xi^2} + \xi \frac{v'(s)}{v(s)} \frac{\partial H}{\partial \xi} + \frac{\partial H}{\partial s} + \psi(\xi, s) H = 0$$

in the strip $0 < s < t$, ξ in R_1.

$H(\xi, s)$ satisfies at each ξ in R_1 the condition

$$\lim_{s \to t-} H(\xi, s) = \sigma(\xi).$$

Furthermore, $\partial^2 H/\partial \xi^2$, $\partial H/\partial \xi$ and $\partial H/\partial s$ are continuous for ξ in R_1, $0 < s < t$.

Let us consider some examples.

Example 1. Wiener process: $u(p) = p$, $v(p) = 1$, $0 \leqslant p \leqslant t$.

$$\frac{1}{2} \frac{\partial^2 G}{\partial \xi^2} + \frac{\partial G}{\partial s} + \psi(\xi, s) G = 0.$$

Example 2. Doob-Kac process: $u(p) = p$, $v(p) = 1 - p$, $0 \leqslant p \leqslant t < 1$.

$$\frac{1}{2} \frac{\partial^2 G}{\partial \xi^2} - \frac{\xi}{1-s} \frac{\partial G}{\partial \xi} + \frac{\partial G}{\partial s} + \psi(\xi, s) G = 0.$$

Example 3. Ornstein-Uhlenbeck family of processes:

$$u(p) = \sigma^2 e^{\alpha p}, v(p) = e^{-\alpha p}, \quad \sigma^2 > 0, \alpha > 0, 0 \leqslant p \leqslant t.$$

$$\sigma^2 \alpha \frac{\partial^2 G}{\partial \xi^2} - \alpha \xi \frac{\partial G}{\partial \xi} + \frac{\partial G}{\partial s} + \psi(\xi, s) G = 0.$$

Our view so far has been to say that a Gaussian Markov integral of a certain functional satisfies a certain partial differential equation. But we will now show that you can essentially turn this around. In other words, if you start with a certain type partial differential equation, you can build an appropriate Gaussian Markov process such that averaging an appropriate functional over its paths solves the partial differential equation.

To be precise, let us assume that we are given the partial differential equation

$$\frac{A(s)}{2} \frac{\partial^2 G}{\partial \xi^2} + \xi B(s) \frac{\partial G}{\partial \xi} + \frac{\partial G}{\partial s} + \psi(\xi, s) G = 0$$

and the boundary condition

$$\lim_{s \to t-} G(\xi, s) = \sigma(\xi)$$

where $A(s) > 0$ on $0 \leqslant s \leqslant t$ for some $t > 0$ and has a continuous derivative on $0 \leqslant s \leqslant t$, $B(s)$ is real valued and has a continuous derivative on $0 \leqslant s \leqslant t$, and the $\psi(y, p)$ and $\sigma(y)$ satisfy the conditions of the previous theorem.

We can now construct a solution to this problem. If we let

$$v(p) = \exp\left\{\int_0^p B(x)\,dx\right\}$$

and
$$u(p) = v(p)\int_0^p A(x)/v^2(x)\,dx, \quad 0 \leqslant p \leqslant t,$$

these functions filfill the requirements for the factorable covariance function. Also $v'(s)/v(s) = B(s)$ and $u'(s)v(s) - u(s)v'(s) = A(s)$. Let the $R(a, b)$ covariance be as in the Theorem and let $\{x(p), s \leqslant p \leqslant t\}$ be the Gaussian Markov stochastic process determined by R and the mean function which is identically zero. Then the integral of the functional of the Theorem for this process satisfies the system.

In 1951 M. Rosenblatt derived some interesting integral and partial differential equations for an n-dimensional Wiener process. [31]. Let

$$x_k(t),\ x_k(0) = 0,\ 0 \leqslant t < \infty \quad (k = 1, \ldots, n)$$

be elements of n independent Wiener processes. Let $\bar{x}(t)$ be an element of the product space of the n independent Wiener spaces. Assume that $V(t, \bar{x})$ is a Borel measurable function of t and the n-dimensional vector $\bar{x} = (x_1, x_2, \ldots, x_n)$ which is bounded in every finite Euclidean sphere of the $(t, \bar{x})$ space, $0 \leqslant t < \infty$. Let

$$Q(t, \bar{x}) = E\left\{\exp\left[-u\int_0^t V(\tau, \bar{x}(\tau))\,d\tau\right] \mid \bar{x}(t) = \bar{x}\right\} \frac{\exp(-\bar{x}\cdot\bar{x}/(2t))}{(2\pi t)^{n/2}}$$

for $u \geqslant 0$. Then $Q(t, \bar{x})$ satisfies the integral equation

$$Q(t, \bar{x}) + u \int_0^t \int_{-\infty}^{\infty} \overset{(n)}{\cdots} \int_{-\infty}^{\infty} C(\bar{\xi}, \tau; \bar{x}, t) V(\tau, \bar{\xi}) Q(\tau, \bar{\xi})\,d\bar{\xi}\,d\tau = C(\bar{0}, 0; \bar{x}, t)$$

where
$$C(\bar{\xi}, \tau; \bar{x}, t) = \frac{\exp(-|\bar{x} - \bar{\xi}|^2/[2(t-\tau)])}{[2\pi(t-\tau)]^{n/2}}.$$

He then considered n-dimensional forms of partial differential equations similar to those discussed by Cameron and Kac. As an example, he computed the probability density of diffusion from $\bar{0}$ to a point $\bar{x}$, when there is an absorbing barrier at the sphere of radius b about $\bar{0}$.

A pair of integral equations for an appropriate Markov integral were derived by D. A. Darling and A. J. F. Siegert. [11]. Let $x(t)$ be an n-

dimensional Markov process $x(t) = (x_1(t), x_2(t), ..., x_n(t))$. They assume that a function $\Phi(x, t)$ is continuous in t and Borel measurable in x and such that

$$\int_s^t E\{|\Phi(x(\tau), \tau)| \ \ |x(s) = x, x(t) = y\} \, d\tau < \infty.$$

The transition probability for $x(t)$ is

$$P(x, s; y, t) = \text{Prob } \{x(t) \leqslant y | x(s) = x\}.$$

Let

$$R(x, s | y, t) = E\left\{\exp\left[i\xi \int_s^t \Phi(x(\tau), \tau) \, d\tau\right] | x(s) = x, x(t) \leqslant y\right\} P(x, s; y, t).$$

This function satisfies the pair of integral equations

(1) $R(x, s | y, t) = P(x, s; y, t) + i\xi \int_s^t \int_{-\infty}^\infty \overset{(n)}{\cdots} \int_{-\infty}^\infty d_\alpha R(x, s | \alpha, \tau)$
$\times \Phi(\alpha, \tau) P(\alpha, \tau; y, t) \, d\tau$

(2) $R(x, s | y, t) = P(x, s; y, t) + i\xi \int_s^t \int_{-\infty}^\infty \overset{(n)}{\cdots} \int_{-\infty}^\infty d_\alpha P(x, s; \alpha, \tau)$
$\times \Phi(\alpha, \tau) R(\alpha, \tau | y, t) \, d\tau.$

It is instructive to show their proof of the first equation. One starts with the elementary identity

$$\exp\left\{i\xi \int_s^t \Phi(x(\tau), \tau) \, d\tau\right\} \equiv 1 + i\xi \int_s^t \Phi(x(u), u) \exp\left\{i\xi \int_s^u \Phi(x(\tau), \tau) \, d\tau\right\} du.$$

Multiply both sides by $P(x, s; y, t)$ and take conditional expectations under the condition $x(s) = x$, $x(t) \leqslant y$. The left-side becomes $R(x, s | y, t)$ and the expectation of the right side (taken under the integral $\int_s^t$, as justified by the condition on Φ) becomes

$P(x, s; y, t) + i\xi \int_s^t E\left\{\Phi(x(u), u) \exp\left[i\xi \int_s^u \Phi(x(\tau), \tau) \, d\tau\right] | x(s) = x, x(t) \leqslant y\right\}$
$\times du \, P(x, s; y, t)$

$= P(x, s; y, t) + i\xi \int_s^t \int_{-\infty}^\infty \overset{(n)}{\cdots} \int_{-\infty}^\infty E\Big\{\Phi(x(u), u)$
$\times \exp\left[i\xi \int_s^u \Phi(x(\tau), \tau) \, d\tau\right] | x(s) = x, x(u) = \alpha, x(t) \leqslant y\Big\}$
$\times d_\alpha P(x, s; \alpha, u) P(\alpha, u; y, t) \, du$

by the definition of conditional expectation

$$= P(x, s; y, t) + i\xi \int_s^t \int_{-\infty}^\infty \overset{(n)}{\ldots} \int_{-\infty}^\infty \Phi(\alpha, u) E\left\{\exp\left[i\xi \int_s^u \Phi(x(\tau), \tau)\,d\tau\right]\Big| x(s)\right.$$

$$= x,\; x(u) = \alpha\bigg\} d_\alpha P(x, s; \alpha, u)\, P(\alpha, u; y, t)\, du$$

$$= P(x, s; y, t) + i\xi \int_s^t \int_{-\infty}^\infty \overset{(n)}{\ldots} \int_{-\infty}^\infty d_\alpha R(x, s\,|\,\alpha, u)\, \Phi(\alpha, u)\, P(\alpha, u; y, t)\, du.$$

The proof of equation (2) starts with the identity

$$\exp\left\{i\xi \int_s^t \Phi(x(\tau), \tau)\, d\tau\right\} = 1 + i\xi \int_s^t \Phi(x(u), u) \exp\left[i\xi \int_u^t \Phi(x(\tau), \tau)\, d\tau\right] du.$$

To derive one of the integral equations in [2], the author found it convenient to put $i\xi \Phi(\alpha, u) = \psi(\alpha, u)$, to multiply both sides of the second identity by $\sigma\{x(t)\}$, perform a similar analysis to that for equation (1), then put $y = +\infty$ to obtain

$$(3)\quad E\left\{\exp\left[\int_s^t \psi(x(\tau), \tau)\, d\tau\right] \sigma(x(t))\,\Big|\, x(s) = x\right\}$$

$$= \int_{-\infty}^\infty \sigma(y)\, p(x, s; y, t)\, dy + \int_s^t \int_{-\infty}^\infty \psi(\alpha, u)\, E$$

$$\times \left\{\exp\left[\int_u^t \psi(x(\tau), \tau)\, d\tau\right] \sigma(x(t))\,\Big|\, x(u) = \alpha\right\} p(x, s; \alpha, u)\, d\alpha\, du.$$

It was assumed that $\sigma(y)$ was a complex valued, integrable function of y, with a continuous, bounded first derivative for y in $(-\infty, \infty)$.

In the Gaussian Markov case, the transition probability has a density $p(x, s; y, t)$ and the corresponding function

$$r(x, s\,|\,y, t) = E\left\{\exp\left[\int_s^t \theta[x(\tau), \tau]\, d\tau\right]\Big|\, x(s) = x,\, x(t) = y\right\} p(x, s; y, t)$$

satisfies a similar pair of integral equations. Using this pair, the following theorem was proved:

THEOREM 6. Assume that $\theta(y, \tau)$ and $\theta_y(y, \tau)$ are bounded and continuous complex valued functions in S: $0 \leqslant \tau \leqslant T < \infty$, $y \in \{(-\infty, \infty) - E\}$ where E is a finite set of points.

Then the function $r(x,s|y,t)$ *uniquely* satisfies (as a function of y and t)

(4) $\quad \dfrac{v(t)u'(t)-u(t)v'(t)}{2}\dfrac{\partial^2 r}{\partial y^2}-\dfrac{v'(t)}{v(t)}\dfrac{\partial [yr]}{\partial y}+\theta(y,t)r=\dfrac{\partial r}{\partial t},$

for $\qquad\qquad\qquad 0\leqslant s<t<T,\quad y\in\{(-\infty,\infty)-E\}$

(5) $\quad\lim\limits_{t\to s+}\displaystyle\int_{-\infty}^{\infty} g(x)r(x,s|y,t)\,dx = g(y)\quad$ for every bounded continuous g

(6) $\quad\lim\limits_{|y|\to\infty} r(x,s|y,t)=0$

(7) $\quad\dfrac{\partial r}{\partial y},\dfrac{\partial^2 r}{\partial y^2},\dfrac{\partial r}{\partial t}\quad$ continuous for $0\leqslant s<t<T$,

$\qquad\qquad y\in\{(-\infty,\infty)-E\}\cdot\dfrac{\partial r}{\partial y}\quad$ is also continuous if $y\in E$.

Furthermore, $r(x,s|y,t)$ uniquely satisfies (as a function of x and s):

(8) $\quad \dfrac{[v(s)u'(s)-u(s)v'(s)]}{2}\dfrac{\partial^2 r}{\partial x^2}+x\dfrac{v'(s)}{v(s)}\dfrac{\partial r}{\partial x}+\theta(x,s)r=-\dfrac{\partial r}{\partial s}$

for $\qquad\qquad\qquad 0<s<t\leqslant T,\quad x\in\{(-\infty,\infty)-E\}.$

(9) $\quad\lim\limits_{s\to t-}\displaystyle\int_{-\infty}^{\infty} g(y)r(x,s|y,t)\,dy = g(x)\quad$ for every continuous bounded g.

(10) $\quad\lim\limits_{|x|\to\infty} r(x,s|y,t)=0.$

(11) $\quad\dfrac{\partial r}{\partial x},\dfrac{\partial^2 r}{\partial x^2},\dfrac{\partial r}{\partial s}\quad$ are continuous for $0<s<t\leqslant T$,

$\qquad\qquad x\in\{(-\infty,\infty)-E\}\cdot\dfrac{\partial r}{\partial x}\quad$ is also continuous if $x\in E$.

Conditions (5) and (9) state in a precise fashion that the r function acts like a Dirac delta function. That "function" is used by physicists rather extensively. It is used to idealize instantaneous forces, and may be thought of as a "function" which vanishes except at one point, but yet possesses the property that $\int_{-\infty}^{\infty}\delta(t)\,dt=1$. A Dirac delta "function" should possess the "sifting property" which we have used above, namely $\int_{-\infty}^{\infty}\delta(t)f(t)\,dt=f(0)$ for every continuous, bounded function f. In a technical sense, a Dirac delta function does not satisfy the usual definition of a function, but still it is a convenient aid to one's thinking.

These equations and conditions were used as building blocks for Green's functions for generalized Schroedinger equations (See [3] and [4]).

Once again, if one is given equations (4) and (8) and the coefficients are slightly restricted, one can construct a Gaussian Markov process and the allied integral satisfies (4)–(11). See [2], page 31.

The same paper contained a theorem which proved quite useful in calculating certain distributions of random walk for the Ornstein-Uhlenbeck process.

THEOREM 7. Assume that $\{x(t), 0 \leqslant t < \infty\}$ is as before, and has a stationary transition density function. Let $V(y)$ be a non-negative function for which $V(y)$ and $V'(y)$ are continuous and bounded, $y \notin E$, E a finite set of points. Let

$$\hat{Q}(y) = \int_0^\infty e^{-\lambda t} r(0, 0 | y, t) \, dt, \quad \lambda > 0,$$

where

$$r(0, 0 | y, t) = E\left\{\exp\left[-u \int_0^t V[x(\tau)] \, d\tau\right] \bigg| x(0) = 0, \quad x(t) = y\right\} p(0, 0; y, t).$$

Then $\hat{Q}$ is the unique solution of

(1) $A \dfrac{d^2 \hat{Q}}{dy^2} - B \dfrac{d}{dy}(y\hat{Q}) - uV(y)\hat{Q} = \lambda \hat{Q}$ for $y \neq 0, y \in E$

where $A(>0)$ and B are the constants in the forward diffusion equation. $\hat{Q}$ is subject to the following conditions:

(2) $\hat{Q}'(0+) - \hat{Q}'(0-) = -1/A$

(3) $\lim\limits_{|y| \to \infty} \sqrt{|y|} \, \hat{Q}(y) = 0$

(4) $\hat{Q}(y)$ continuous for all y

(5) $\hat{Q}'(y)$ continuous for $y \neq 0$

(6) $\hat{Q}''(y)$ continuous for $y \neq 0$, $y \notin E$

(7) $|\hat{Q}'(y)| < C + D/\lambda$, C and D constants, $y \neq 0$.

For the Ornstein-Uhlenbeck process, let

$$P \equiv P[-b < \inf_{0 \leqslant s \leqslant t} x(s) \leqslant \sup_{0 \leqslant s \leqslant t} x(s) < a \,|\, x(0) = 0]$$

where $b>0$, $a>0$. The continuity of the sample functions allows P to be expressed as follows:

$$P = \lim_{u\to\infty} E\left\{\exp\left[\int_0^t \theta_u[x(s)]\,ds\right]\Big| x(0)=0\right\}$$

where
$$\theta_u(y) = \begin{cases} 0, & -b < y < a \\ -u, & y < -b, y > a. \end{cases}$$

This is the same as to say

$$P = \lim_{u\to\infty} \int_{-\infty}^{\infty} r(0,0\,|\,y,t)\,dy$$

where r satisfies the partial differential equation and conditions and $E = \{-b, a\}$. By the previous theorem, to obtain P we can solve

$$\sigma^2\beta \frac{d^2\hat{Q}}{dy^2} + \beta \frac{d(y\hat{Q})}{dy} + [-uV(y) - \lambda]\hat{Q} = 0,\ y \notin \{-b, 0, a\}$$

with
$$V(y) = \begin{cases} 0, & -b < y < a,\ a > 0,\ b > 0 \\ 1, & y < -b, y > a \end{cases}$$

where $\hat{Q}$ is subject to conditions (2) through (7).

For simplicity we let $\sigma^2 = 1$ but retain β because it may be interpreted as the reciprocal of the friction coefficient, and for large values of β and t, the Ornstein-Uhlenbeck process serves as a model for the Brownian motion of an elastically bound particle. See [23].

With some effort the above system is solved and yields an expression for the Laplace transform of P in terms of Weber functions. See [2] for details. The expression includes as subcases the Laplace transforms for the distribution functions

$$P[\sup_{0 \leqslant s \leqslant t} x(s) < a\,|\,x(0)=0]$$

and
$$P[\max_{0 \leqslant s \leqslant t} |x(s)| < a\,|\,x(0)=0].$$

An interesting paper about the title of this subsection is by M. D. Donsker and J. L. Lions [12].

Exercise

If $g(x)$ is bounded and continuous on $(-\infty, \infty)$, prove that

$$\lim_{\sigma^2 \to 0} \int_{-\infty}^{\infty} g(x) \frac{e^{-x^2/(2\sigma^2)}}{\sqrt{2\pi\sigma^2}}\,dx = g(0).$$

Thus the normal probability density with mean 0 and variance σ^2 acts as a Dirac delta function as $\sigma^2 \to 0$.

4.4. Relation between Wiener and Gaussian Markov integrals

Reference [13] by J. L. Doob showed how to transform a Gaussian Markov process into the Wiener process.

Using that idea and a paper by D. Varberg [35], the following result was proved in [1]:

Theorem 8. Let $\{x(p), s \leqslant p \leqslant t\}$ be a Gaussian Markov process with mean function

$$m(p) = \xi v(p)/v(s), \quad s \leqslant p \leqslant t \quad \text{(for } \xi \text{ real) and covariance function}$$

$$R(a, b) = \begin{cases} U(a) v(b), & a \leqslant b \\ U(b) v(a), & b \leqslant a \end{cases}$$

where $\qquad U(p) = u(p) - u(s) v(p)/v(s), \quad s \leqslant p \leqslant t.$

Then for all (B) measurable functions F,

$$\int_{C_\xi[s,\,t]} F[x]\,d_{m,R}x = \int_{C[s,\,t]} F[y(\cdot) + \xi v(\cdot)/v(s)]\,d_R y$$

$$= \int_{C[0,\,U(t)/v(t)]}^{W} F[v(\cdot)x(U(\cdot)/v(\cdot)) + \xi v(\cdot)/v(s)]\,d_w x.$$

There is useful corollary, due to E. Cuthill [10] which says the above is equal to

$$\int_{C[0,\,1]}^{W} F\left[v(\cdot)\sqrt{\frac{U(t)}{v(t)}} x\left(\frac{U(\cdot)/v(\cdot)}{U(t)/v(t)}\right) + \xi \frac{v(\cdot)}{v(s)}\right] d_w x.$$

An interesting subcase is that

$$\int_{C[s,\,t]} F(y)\,d_w y = \int_{C[0,\,1]} F\left[\sqrt{t-s}\, X\left(\frac{(\cdot)-s}{t-s}\right)\right] d_w x.$$

Thus, the Gaussian Markov integral of a simple functional may be easy to interpret, but it may be easier to calculate the Wiener integral of the more complicated functional.

Mention should also be made of another representation of the Doob-Kac process contained on page 65 of the book [5] by P. Billingsley. Let

$\{W_t, 0 \leq t \leq 1\}$ and $\{W_t^0, 0 \leq t \leq 1\}$ be the Wiener and Doob-Kac processes, respectively. Then $W_t^0 = W_t - tW_1$, $0 \leq t \leq 1$. Billingley refers to the Doob-Kac process as the Brownian bridge or tied-down Brownian motion, motivated by the fact that $W_0^0 = W_1^0 = 0$ with probability one.

4.5. Monte-Carlo approximation of conditional Wiener integrals

One of the most facinating ideas in the area of function space integrals is their approximation by Monte Carlo techniques. It should be clear to the reader that function space integrals can be difficult to evaluate. Monte Carlo approximations considerably enlarge the class of functionals to which a numerical value can be given to their integrals. References [18] and [19] are by Lloyd Fosdick; reference [20] is by Lloyd Fosdick and Harry Jordan. They refer to the conditioned Wiener process or the Doob-Kac process. They relate to the connection between certain Wiener integrals and the Schroedinger equation, which will be discussed in detail in a later chapter. Papers [3] and [4] show how to convert a conditioned Gaussian Markov integral to one amenable to their techniques.

In [18] Fosdick starts by using a theorem due to Cameron [8] which allows an approximation of the Wiener integral by an n-dimensional Riemann integral. Such an approximation approaches the Wiener integral as $n \to \infty$. He then estimates the value of the n-dimensional Riemann integral by a Monte Carlo sampling technique. References [19] and [20] use alternate approaches to the approximation of the Wiener integral, rather than Cameron's Simpson's Rule.

We will now describe the methods of [18] based on pages 1252–1253 of that reference. We start with one of Cameron's "rectangular approximations to Wiener integrals". See [8].

THEOREM. Let $F[x(\cdot)]$ be continuous in the Hilbert topology on the space $C_0[0, 1]$ and let

$$|F[x(\cdot)]| \leq H\left(\int_0^1 x^2(\tau)\,d\tau\right) \quad \text{on } C_0[0, 1]$$

where $H(u)$ is monotonically increasing and the Wiener integral $E\{H\}$ satisfies $E\{H\} < \infty$.

Let $\alpha_1(\tau), \alpha_2(\tau), \ldots$ be a complete orthonormal set on the interval $0 \leq \tau \leq 1$ such that $\alpha_j(\tau) \in C_0[0, 1]$, $j = 1, 2, \ldots$ and let $\beta_1(\tau), \beta_2(\tau), \ldots$ be obtained by applying the Schmidt orthogonalization process to the sequence of integrals

$$\int_\tau^1 \alpha_1(s)\, ds, \quad \int_\tau^1 \alpha_2(s)\, ds, \ldots$$

Let the relation between these integrals and the β_k's be given by

$$\int_\tau^1 \alpha_k(s)\, ds = \sum_{j=1}^k \gamma_{jk} \beta_j(\tau).$$

Then the Wiener integral of $F[x(\,\cdot\,)]$ is given by the formula

$$E\{F\} = \lim_{n\to\infty} \int_{-\infty}^\infty \cdots \int_{-\infty}^\infty F\left[\sum_{k=1}^n \sum_{j=1}^k \gamma_{j,k}\, \xi_j\, \alpha_k(\,\cdot\,)\right] \pi^{-n/2}$$

$$\times \exp[-\xi_1^2 - \xi_2^2 - \ldots - \xi_n^2]\, d\xi_1 \ldots d\xi_n.$$

Remember that $F[x(\,\cdot\,)]$ is continuous in the Hilbert topology on $C_0[0,1]$ if for $x_0 \in C_0[0,1]$ and $\varepsilon > 0$, there exists $\delta > 0$ such that if $x(\tau) \in C_0[0,1]$ and such that

$$\left[\int_0^1 |x(\tau) - x_0(\tau)|^2\, d\tau\right]^{\frac{1}{2}} < \delta$$

then $\quad |F[x(\,\cdot\,)] - F[x_0(\,\cdot\,)]| < \varepsilon.$

Now let the orthonormal set

$$\alpha_j(\tau) = \sqrt{2}\,\sin(j - \tfrac{1}{2})\pi\tau.$$

Then

$$\int_\tau^1 \alpha_j(s)\, ds = \frac{\sqrt{2}\,\cos(j - \tfrac{1}{2})\pi\tau}{(j - \tfrac{1}{2})\pi}$$

and

$$\gamma_{jk} = \delta_{jk}[(j - \tfrac{1}{2})\pi]^{-1},$$

where δ_{jk} is the Kronecker delta.

Hence Cameron's theorem gives

$$E\{F\} = \lim_{n\to\infty} \int_{-\infty}^\infty \cdots \int_{-\infty}^\infty F[x_n(\,\cdot\,)]\, \pi^{-n/2} \exp[-\xi_1^2 - \xi_2^2 - \ldots - \xi_n^2]\, d\xi_1 \ldots d\xi_n,$$

where

$$x_n(\tau) = \sum_{j=1}^n \xi_j \frac{\sqrt{2}\,\sin(j - \tfrac{1}{2})\pi\tau}{(j - \tfrac{1}{2})\pi}$$

The lack of 2's in the normal densities is due to a different choice of constant in the Wiener covariance kernel.

By the rectangle approximation to the Wiener integral $E\{F\}$ we shall mean

$$E\{F\}_n = \int_{-\infty}^{\infty} \cdots \int_{-\infty}^{\infty} F[x_n(\cdot)]$$

(1)
$$\pi^{-n/2} \exp[-\xi_1^2 - \xi_2^2 - \cdots - \xi_n^2] d\xi_1 \ldots d\xi_n.$$

The n-dimensional integral in (1) is now evaluated by Monte Carlo sampling. In this sampling method a set k of n numbers is randomly selected from a Gaussian distribution with mean 0 and variance $\tfrac{1}{2}$. This set is denoted by $\xi_{1,k}, \xi_{2,k}, \ldots, \xi_{n,k}$.

For this set we evaluate

$$x_{n,k} = \sum_{j=1}^{n} \frac{\xi_{j,k}\sqrt{2}\sin(j-\tfrac{1}{2})\pi\tau}{(j-\tfrac{1}{2})\pi}$$

and then the functional $F[x_{n,k}(\cdot)]$.

Repeat this process R times and form the quantity

$$E\{F\}_{n,R} = \frac{1}{R} \sum_{k=1}^{R} F[x_{n,k}(\cdot)]$$

This is the second stage of the approximation. By the law of large numbers $E\{F\}_{n,R} \to E\{F\}_n$ as $R \to \infty$, in probability, i.e.,

$$\Pr\{|E\{F\}_{n,R} - E\{F\}_n| > \varepsilon\} \to 0 \quad \text{for every } \varepsilon > 0 \text{ as } R \to \infty.$$

4.6. Radon-Nikodym derivatives of Gaussian processes

Consider two Gaussian probability measures P_ϱ and P_r determined by covariance functions $\varrho(s,t)$ and $r(s,t)$, respectively (the mean functions will be assumed to be identically zero). In [15], Jacob Feldman showed that P_ϱ and P_r are either equivalent or perpendicular (mutually singular). We say that P_ϱ and P_r are equivalent if (1) $P_\varrho(E) = 0$ for every measurable set E for which $P_r(E) = 0$ and (2) $P_r(F) = 0$ for every measurable set F for which $P_\varrho(F) = 0$. To define perpendicularity, let $C[a,b]$ be the basic space. Then P_ϱ and P_r are mutually singular if there exist two disjoint sets A and B whose union is $C[s,t]$ such that, for every measurable set E, $A \cap E$ and $B \cap E$ are measurable and $P_\varrho(A \cap E) = P_r(B \cap E) = 0$. If P_ϱ and P_r are equivalent, the Radon-Nikodym derivative $(dP_\varrho/dP_r)(x)$ exists and is the exponential of a quadratic form in x. If P_r is Wiener measure, Lawrence A.

Shepp [32], p. 352, has shown when this quadratic form may be diagonal, i.e., expressible as $\int_a^b f(t) x^2(t) dt$. Dale Varberg in [36] has considered Gaussian processes for which this quadratic form is diagonal.

The Radon-Nikodym derivative of one Gaussian–Markov measure with respect to another can be used in filtering and prediction problems in electrical engineering. See page 43 of [6], and [30], for example.[1]

Let us now consider Varberg's Theorem 1 of [34] and several of its examples. We need the following definition:

DEFINITION. Let $M[a, b]$ denote the class of all Gaussian processes $\{x(t), a \leqslant t \leqslant b\}$ with mean function identically zero and covariance function $r(s, t)$ given by

$$r(s, t) = \begin{cases} u(s) v(t), & s \leqslant t \\ u(t) v(s), & s \geqslant t \end{cases}$$

where moreover

(1) $u(a) \geqslant 0$

(2) $v(t) > 0$ on $[a, b]$,

(3) u'' and v'' exist and are continuous on $[a, b]$,

(4) $v(t) u'(t) - u(t) v'(t) > 0$ on $[a, b]$.

THEOREM. Let $\{x(t), a \leqslant t \leqslant b\}$ and $\{y(t), a \leqslant t \leqslant b\}$ be two Gaussian processes belonging to $M[a, b]$ with probability measures m_r and m_ϱ determined by their respective covariance functions $r(s, t)$ and $\varrho(s, t)$.

Let

$$r(s, t) = \begin{cases} u(s) v(t), & s \leqslant t \\ u(t) v(s), & t \leqslant s \end{cases},$$

$$\varrho(s, t) = \begin{cases} \theta(s) \Phi(t), & s \leqslant t \\ \theta(t) \Phi(s), & t \leqslant s \end{cases}.$$

Then necessary and sufficient conditions that m_r be equivalent to m_ϱ are that

(5) $v(t) u'(t) - u(t) v'(t) = \Phi(t) \theta'(t) - \theta(t) \Phi'(t)$ on $[a, b]$,

---

[1] A recent reference which relates to these problems, and contains many references is Kailath, Thomas (1971). The Structure of Radon-Nikodym Derivatives with respect to Wiener and Related Measures. *Annals of Math. Statistics* 42, 1054–1067.

(6) $u(a)$ and $\theta(a)$ are either both zero or both non-zero. Moreover, if these conditions are satisfied, the Radon–Nikodym derivative of m_ϱ with respect to m_r is given by

$$\frac{dm_\varrho}{dm_r} = C_1 \exp\left\{\tfrac{1}{2}\left[C_2 x^2(a) + \int_a^b f(t) \, d\left\{\frac{x^2(t)}{\Phi(t) \, v(t)}\right\}\right]\right\},$$

where

$$C_1 = \begin{cases} \{[\Phi(a)v(b)]/[\Phi(b)v(a)]\}^{\frac{1}{2}} & \text{if } \theta(a) = 0 \\ [u(a)v(b)]/[\theta(a)\Phi(b)]\}^{\frac{1}{2}} & \text{if } \theta(a) \neq 0, \end{cases}$$

$$C_2 = \begin{cases} 0 & \text{if } \theta(a) = 0 \\ [\Phi(a)\theta(a) - u(a)v(a)]/[v(a)\Phi(a)\theta(a)u(a)] & \text{if } \theta(a) \neq 0 \end{cases}$$

and

$$f(t) = [v(t)\Phi'(t) - \Phi(t)v'(t)]/[v(t)u'(t) - u(t)v'(t)].$$

Example 1. Let $\{y(t), 0 \leq t \leq T < 1\}$ be the Doob-Kac process with probability measure m_ϱ determined by the covariance function

$$\varrho(s, t) = \begin{cases} s(1-t), & s \leq t \\ t(1-s), & t \leq s \end{cases}.$$

Let $\{x(t), 0 \leq t \leq T\}$ be the Wiener process with covariance function $w(s, t) = \min(s, t)$. These two processes are equivalent and

(7) $\quad \dfrac{dm_\varrho}{dm_w} = (1-T)^{-\frac{1}{2}} \exp\{-x^2(T)/[2(1-T)]\}.$

Example 2. Let $V[a, b]$ denote the class of Ornstein-Uhlenbeck processes determined by

$$\varrho_{\sigma,\beta}(s, t) = \sigma^2 \exp[-\beta|s-t|] = \begin{cases} \theta(s)\Phi(t), & s \leq t \\ \theta(t)\Phi(s), & t \leq s \end{cases}$$

where $\theta(t) = \sigma^2 \exp(\beta t)$, $\Phi(t) = \exp(-\beta t)$, and $\sigma^2 > 0$ and $\beta > 0$.

Let $\{x(t), 0 \leq t \leq T\}$ and $\{y(t), 0 \leq t \leq T\}$ be two processes belonging to $V[0, T]$ with covariance functions $\varrho_{\sigma_0, \beta_0}$ and $\varrho_{\sigma_1, \beta_1}$. Then m_{σ_0, β_0} is equivalent to m_{σ_1, β_1} if and only if $\sigma_0^2 \beta_0 = \sigma_1^2 \beta_1$. Moreover if this condition is fulfilled and if we let $K = 2\sigma_0^2 \beta_0 = 2\sigma_1^2 \beta_1$, then

(8) $\quad \dfrac{dm_{\sigma_1, \beta_1}}{dm_{\sigma_0, \beta_0}} = \left(\dfrac{\beta_1}{\beta_0}\right)^{\frac{1}{2}} \exp\left\{-\dfrac{1}{2K}\left[(\beta_1 - \beta_0)(x^2(0)\right.\right.$
$\left.\left. + x^2(T) - KT) + (\beta_1^2 - \beta_0^2)\int_0^T x^2(t)\,dt\right]\right\}.$

This result is due to Charlotte Striebel [33].

In [35] Varberg derived Radon-Nikodym derivatives for Gaussian Markov processes under three conditions:

(A) Same covariance, different means
(B) Different covariances, same mean
(D) Different covariances, different means

In [32] L. Shepp gives simple necessary and sufficient conditions on the mean and covariance for a Gaussian measure to be equivalent to Wiener measure. He also considers the equivalence of Gaussian measures. He uses Radon-Nikodym derivatives to calculate an interesting distribution for the Wiener process, and to calculate certain function space integrals.

Let us now consider his Theorem 1. Let L^2 and Q^2 be the spaces of square—integrable functions on $[0, T]$ and $[0, T] \times [0, T]$ respectively. Let $\sigma(K) = \{\lambda : K\varphi = \lambda\varphi \neq 0\}$ be the set of eigen-values of the Hilbert-Schmidt operator K defined by $Kf(u) = \int_0^t K(u, t) f(t) dt$. The equation $K\varphi = \lambda\varphi$ means $\varphi \in L^2$ and $\int_0^T K(t, u) \varphi(u) du = \lambda\varphi(t)$.

THEOREM. $\mu \sim \mu_w$ if and only if there exists a kernel $K \in Q^2$ for which

(9) $R(s, t) = \min(s, t) - \int_0^s \int_0^t K(u, v) du dv,$

and

(10) $1 \notin \sigma(K)$ and a function $k \in L^2$ for which

(11) $m(t) = \int_0^t k(u) du.$

The kernel K is unique and symmetric and is given by

$$K(s, t) = -\frac{\partial^2}{\partial s \partial t} R(s, t)$$

for almost every (s, t). The function k is unique and is given by $k(t) = m'(t)$ for almost every t.

Shepp shows that all the eigenvalues $\lambda_j < 1$ and therefore R is strictly positive-definite. In case

$$R_1(s, t) = \frac{\partial R}{\partial s}(s, t)$$

is continuous for $s \neq t$, (9) becomes $R_1(s, s+) - R_1(s, s-) \equiv 1, 0 < s < T$. The interpretation of this is that R has a fixed discontinuity of unit size in its derivative along $s=t$, which is the same discontinuity that $m(s,t)$ has in its derivative.

On page 351, Shepp shows that for the Doob-Kac process,

(12) $$E\left\{\exp\left[-\tfrac{1}{2}\int_0^t f(t)\,x^2(t)\,dt\right]\right\} = [h(1)]^{\tfrac{1}{2}}$$

where $h(t)$ is the unique solution to

(13) $h''(t) = f(t)\,h(t)$

subject to the conditions

$h(0)=0,\ h'(0)=1,$

provided $h(t) > 0$ in $(0, 1]$.

When the solution $h(t)$ of (12) has a zero in $[0,1]$ then (12) is $+\infty$.

When $f(t) > 0$, (12) is finite and Shepp gives references to earlier derivations of this result.

Exercises

1. Verify formula (7).
2. Verify formula (8).
3. Consider (9), (10), and (11) when μ is for the Doob-Kac process.

4.7. Useful distributions

The following collection of distributions should prove useful to researchers in electrical engineering, and other fields. Many of them relate to statistics and the subject of Chapter 6.

A. The Maximum Functional for the Wiener Process. See [13] and references.

(1) $$P\{\max_{0 \leqslant t \leqslant T} x(t) \geqslant z\} = \frac{2}{\sqrt{2\pi T}} \int_z^\infty e^{-w^2/(2T)}\,dw.$$

This is also the distribution for

$$P\{\min_{0 \leqslant t \leqslant T} x(t) \leqslant -Z\}.$$

In both cases, $Z \geqslant 0$.

B. The Abolute Maximum Functional for the Wiener Process.

(2) $P\{\max_{0 \leqslant t \leqslant 1} |x(t)| \leqslant \alpha\} = \dfrac{4}{\pi} \sum_{k=0}^{\infty} \dfrac{(-1)^k}{2k+1} \exp\left\{-\dfrac{(2k+1)^2 \pi^2}{16\alpha^2}\right\}$

were $\alpha > 0$.

C. The Absolute Maximum Functional for the Wiener Process, Conditioned. See [28], page 49.

(3) $P\{\max_{0 \leqslant t \leqslant 1} |x(t)| \leqslant \alpha,\ \gamma_1 < x(1) < \gamma_2\}$

$= \dfrac{2}{\pi} \sum_{k=0}^{\infty} \dfrac{(-1)^k}{2k+1} \left\{\cos\left[\dfrac{(2k+1)(\gamma_1+\alpha)\pi}{2\alpha}\right] - \cos\left[\dfrac{(2k+1)(\gamma_2+\alpha)\pi}{2\alpha}\right]\right\}$

$\times \exp\left\{-\dfrac{(2k+1)^2 \pi^2}{16\alpha^2}\right\}$ for $\alpha > 0$.

Notice that this reduces to (2) for $\gamma_1 = -\alpha$ and $\gamma_2 = \alpha$.

D. A Linear Bound on Wiener Sample Paths. See [13], page 397.

(4) If $a \geqslant 0$, $b > 0$, then $P\{\text{l.u.b.}_{0 \leqslant t < \infty} [x(t) - (at+b)] \geqslant 0\} = e^{-2ab}$

E. Two Linear Bounds on Wiener Sample Paths. See [13], page 398.
If $a \geqslant 0$, $b > 0$, $\alpha \geqslant 0$, $\beta > 0$, then

(5) $P\{\text{l.u.b.}_{0 \leqslant t < \infty} [x(t) - (at+b)] \geqslant 0$ or $\text{g.l.b.}_{0 \leqslant t < \infty} [x(t) + \alpha t + \beta] \leqslant 0\}$

$= \sum_{m=1}^{\infty} \{\exp[-2[m^2 ab + (m-1)^2 \alpha\beta + m(m-1)(\alpha\beta + ab)]]$

$+ \exp[-2[(m-1)^2 ab + m^2 \alpha\beta + m(m-1)(\alpha\beta + ab)]]$

$- \exp[-2[m^2(ab+\alpha\beta) + m(m-1)\alpha\beta + m(m+1)\alpha\beta]]$

$- \exp[-2[m^2(ab+\alpha\beta) + m(m+1)\alpha\beta + m(m-1)ab]]\}.$

In the case $\alpha = a$, $\beta = b$

(6) $P\left\{\text{l.u.b.}_{0 \leqslant t < \infty} \dfrac{|x(t)|}{at+b} \geqslant 1\right\} = 2 \sum_{m=1}^{\infty} (-1)^{m+1} e^{-2m^2 ab}$

F. See [32], page 348, and its reference to a paper by S. Malmquist, [27].

For $b > 0$,

(7) $P\{x(t) < at+b, 0 \leq t \leq T\}$

$$= N\left[\frac{aT+b}{T^{\frac{1}{2}}}\right] - e^{-2ab} N\left[\frac{aT-b}{T^{\frac{1}{2}}}\right]$$

where
$$N(x) = \frac{1}{\sqrt{2\pi}} \int_{-\infty}^{x} e^{-w^2/2} dw.$$

G. Same as F for Conditioned Wiener Process. See [27], page 526.
For $a \geq 0, b > 0$

(8) $\quad P\{x(t) \leq at+b, \tau_1 \leq t \leq \tau_2 | x(\tau_1) = s_1, x(\tau_2) = s_2\}$

$$= 1 - \exp\left\{\frac{-2(P_1-s_1)(P_2-s_2)}{(1-R)^2 \tau_2}\right\}$$

where $R = [\tau_1/\tau_2]^{\frac{1}{2}}$ and $s_1 \leq P_1 = ax+b$ and $s_2 \leq P_2 = ay+b$.

Malmquist also generalizes formula (5) to allow a conditioning.

H. Arc Sine Law for the Wiener Process.
Let $A_x =$ Lebesque measure of those t in [0, 1] for which $x(t) > 0$.

(9) $\quad P[x(\cdot): A_x \leq u] = \frac{2}{\pi} \arcsin \sqrt{u} \quad$ for $0 < u < 1$.

This result is due to Paul Lévy.

I. Related to G.
Let $T =$ supremum of those t in [0, 1] for which $x(t) = 0$.

(10) Then $\quad P[T \leq t] = \frac{2}{\pi} \arcsin \sqrt{t} \quad$ for $0 < t < 1$.

See [5], page 83.

J. The maximum Functional for the Doob-Kac Process. See [13], page 402.

(11) $P[\max_{0 \leq t \leq 1} x(t) < b] = 1 - e^{-2b^2}, \quad b > 0.$

This is also the distribution of

$$P[\min_{0 \leq t \leq 1} x(t) > -b].$$

K. The Absolute Maximum Functional for the Doob-Kac Process. See [13], page 402.

(12) $P[\max_{0 \leq t \leq 1} |x(t)| < b] = 1 - 2 \sum_{m=1}^{\infty} (-1)^{m+1} e^{-2m^2 b^2}$

The "2" before the summation is missing in the above reference. It is present in a related earlier paper by W. Feller. See [16], page 178.

L. A Band for the Doob-Kac Process. See [13], page 403.

(13) $P[\min_{0 \leq t \leq 1} x(t) \geq -\lambda_1, \max_{0 \leq t \leq 1} x(t) \leq \lambda_2]$

$= 1 - \sum_{m=1}^{\infty} \{\exp[-2[m\lambda_2 + (m-1)\lambda_1]^2]$

$+ \exp[-2[(m-1)\lambda_2 + m\lambda_1]^2] - 2\exp[-2m^2(\lambda_1 + \lambda_2)^2]\}$.

M. Related to H; Doob-Kac Process.

Let A_x be the Lebesque measure of those t in $[0, 1]$ for which $x(t) > 0$.

(14) $P\{x(\cdot): A_x \leq \alpha\} = \alpha$, $0 < \alpha < 1$.

See [5], page 86.

N. Linear Bounds on Doob-Kac Sample Paths.
If $\alpha > 0$, $\beta > 0$

(15) $P[x(t) < \alpha(1-t) + \beta t, 0 \leq t \leq 1] = 1 - e^{-2\alpha\beta}$

and

(16) $P[-\alpha(1-t) - \beta t < x(t) < \alpha(1-t) + \beta t, 0 \leq t \leq 1]$

$= 1 - 2 \sum_{m=1}^{\infty} (-1)^{m+1} \exp\{-2m^2 \alpha \beta\}$

This is based on [13] but in this form will be found on page 182 of [21].

Exercises

1. Compute $P\{\max_{0 \leq t \leq 1} X(t) \geq 1\}$ for the Wiener process.
2. Compute $P\{\max_{0 \leq t \leq 1} |X(t)| \leq 1\}$ in distribution B accurate to .01. How many terms did it take?
3. Graphically interpret the set of functions in distribution D, and illustrate with $a = b = .5$.
4. Compute $P\{X(t) < t + .5, 0 \leq t \leq 2\}$ in distribution F.
5. Compute $P\{X(\cdot): A_x \leq f\}$ for $f = .25, .50, .75$, in distribution H.

REFERENCES

1. BEEKMAN, J. A. (1965). Gaussian processes and generalized Schroedinger equations. *J. Math. Mech.* **14**, 789–806.
2. — (1967). Gaussian Markov processes and a boundary value problem. *Trans. Amer. Math. Soc.* **126**, 29–42.
3. — (1969). Green's functions for generalized Schroedinger equations. *Nagoya Math. J.* **35**, 133–150. Correction: *Nagoya Math. J.* **39** (1970), 199.
4. — (1971). Sequential Gaussian Markov Integrals. *Nagoya Math. J.* **42**, 9–21.
5. BILLINGSLEY, P. (1968). *Convergence of Probability Measures.* Wiley, New York.
5. BUCY, R. and JOSEPH, P. (1968). *Filtering for Stochastic Processes with Applications to Guidance.* Wiley, New York.
7. CAMERON, R. H. and MARTIN, W. T. (1944). Transformations of Wiener integrals under translations. *Ann. Math.* **45**, 386–396.
8. CAMERON, R. H. (1951). A 'Simpson's Rule' for the numerical evaluation of Wiener's integrals in function space. *Duke Math. J.* **18**, 111–130.
9. — (1954). The generalized heat flow equation and a corresponding Poisson formula. *Ann. Math.* **59**, 434–462.
10. CUTHILL, E. (1953). Integrals on spaces of functions which are real and continuous on finite and infinite intervals. Thesis, Univ. of Minnesota.
11. DARLING, D. A. and SIEGERT, A. J. F. (1955). On the distribution of certain functionals of Markoff processes. Rand Report, P-429, Rand Corporation, Santa Monica, California.
12. DONSKER, M. D. and LIONS, J. L. (1962). Fréchet-Volterra variational equations, boundary value problems, and function space integrals. *Acta Mathematica* **108**, 142–228.
13. DOOB, J. L. (1949). Heuristic approach to the Kolmogorov-Smirnov theorems. *Ann. Math. Stat.* **20**, 393–403.
14. — (1953). *Stochastic Processes.* Wiley, New York.
15. FELDMAN, J. (1958). Equivalence and perpendicularity of Gaussian processes. *Pacific J. Math.* **8**, 699–708.
16. FELLER, W. (1948). On the Kolmogorov-Smirnov limit theorems for empirical distributions. *Ann. Math. Stat.* **19**, 177–189.
17. FEYNMAN, R. P. (1948). Space-time approach to non-relativistic quantum mechanics. *Rev. Mod. Phys.* **20**, 367–387.
18. FOSDICK, L. D. (1962). Numerical estimation of the partition function in quantum statistics. *J. Math. Physics* **3**, 1251–1264.
19. — (1965). Approximation of a class of Wiener integrals. *Math. of Computation* **19**, 225–233.

20. FOSDICK, L. D. and JORDAN, H. F. (1968). Approximation of a conditional Wiener integral. *J. Computational Physics* **3**, 1–16.
21. HAJEK, J. and Z. SIDAK (1967). *Theory of Rank Tests*. Academic Press, New York.
22. KAC, M. and SIEGERT, A. J. F. (1947). On the theory of noise in radio receivers with square law detectors. *J. Applied Physics* **18**, 382–397.
23. KAC, M. (1947). Random walk and the theory of Brownian motion. *Amer. Math. Monthly* **54**, 369–391.
24. — (1949). On distributions of certain Wiener functionals. *Trans. Amer. Math. Soc.* **65**, 1–13.
25. — (1949). On deviations between theoretical and empirical distributions. *Proc. Natl. Acad. Sci. U.S.A.* **35**, 252–257.
26. — (1951). On some connections between probability theory and differential and integral equations. *Proc. Second Berkeley Symposium*, pp. 189–215. Univ. of Calif. Press, Berkeley, Calif.
27. MALMQUIST, S. (1954). On certain confidence contours for distribution functions. *Ann. Math. Stat.* **25**, 523–533.
28. NELSON, E. O. (1959). A solution of the generalized heat flow equation in a bounded region as a Wiener integral. Ph. D. Thesis, Univ. of Minnesota, Minneapolis.
29. ORNSTEIN, L. S. and UHLENBECK, G. E. (1930). On the theory of the Brownian motion. *Physics Rev.* **36**, 823–841.
30. PARZEN, E. (1963). Probability density functionals and reproducing kernel Hilbert spaces, *Symposium on Time Series Analysis*, edited by M. Rosenblatt. Wiley, New York.
31. ROSENBLATT, M. (1951). On a class of Markov processes. *Trans. Amer. Math. Soc.* **71**, 120–135.
32. SHEPP, L. A. (1966). Radon-Nikodym derivatives of Gaussian measures. *Ann. Math. Stat.* **37**, 321–354.
33. STRIEBEL, C. T. (1959). Densities for stochastic processes. *Ann. Math. Stat.* **30**, 559–567.
34. VARBERG, DALE (1961). On equivalence of Gaussian measures. *Pacific J. Math.* **11**, 751–762.
35. — (1962). Gaussian measures and a theorem of T. S. Pitcher. *Proc. Amer. Math. Soc.* **13**, 799–807.
36. — (1967). Equivalent Gaussian measures with a particularly simple Radon-Nikodym derivative. *Ann. Math. Stat.* **38**, 1027–1030.
37. WIENER, N. (1930). Generalized harmonic analysis. *Acta Mathematica* **55**, 117–258, especially pages 214–234.

CHAPTER 5

CONNECTIONS BETWEEN THE TWO PROCESSES

In this brief chapter, seven references will be mentioned.

Chapter VIII of J. L. Doob's book on stochastic processes, [3], discusses processes with independent increments. It contains the following intriguing idea. Let $\{x_t, t \in T\}$ be a centered process with stationary independent increments. The concept of centering is discussed on pages 407–417 of [3]. Then there exist processes $x_t^{(1)}, ..., x_t^{(n)}, y_t$ which are mutually independent such that: the $x_t^{(j)}$ process is a Poisson process with average occurrence rate C_j; the y_t process is a Brownian motion process with variance parameter 1; $\lambda_1, ..., \lambda_n, C_1, ..., C_n, \sigma, \beta$ are constants; $\sigma \geq 0$; the λ_j's are distinct and none vanish; and for $t \in T$,

$$x_t = \sum_{j=1}^{n} \lambda_j x_t^{(j)} + \sigma y_t + \beta t.$$

Reference [7] by Donald L. Iglehart uses distributions for the Wiener process to approximate certain distributions in collective risk theory. He defines a sequence of risk reserve processes $\{X_n(t), t \geq 0\}$ $n = 1, 2, ...$ and shows that these processes converge weakly to a Brownian motion process with a drift. It relies heavily on the theory of weak convergence of probability measures on function spaces as explained in Billingsley's book mentioned earlier.

A sequence $\{X_n\}$ of random variables converges weakly to the random variable X if the sequence $\{F_n\}$ of distribution functions corresponding to $\{X_n\}$ converges weakly to the distribution function F of X, i.e. if $F_n(x) \to F(x)$ for all continuity points of F. These concepts are denoted by $X_n \Rightarrow X$ and $F_n \xrightarrow{D} F$. A valuable result of this definition of weak convergence is that if $g(x)$ is a measurable function and the set D_g of its points of discontinuity has zero probability with respect to F, then $X_n \Rightarrow X$ implies that $g(X_n) \Rightarrow g(X)$. The ideas of weak convergence can easily be extended from the real line to any finite Euclidean space.

In fact, we can extend the ideas to metric spaces.

DEFINITIONS. A metric space is a set S and a function d defined on pairs of points of S such that:

(1) $d(x, y) = 0$ if and only if $x = y$;

(2) $d(x, y) = d(y, x)$;

(3) $d(x, y) + d(y, z) \geq d(x, z)$.

A neighborhood of a point $x \in S$ is a set $N(x; r) = \{y; d(x, y) < r\}$ for $r > 0$. A sequence $\{x_n\}$ of points in S converges to $x \in S$ if

$$\lim_{n \to \infty} d(x_n, x) = 0.$$

The metric spaces of interest in this chapter and the next are $C_0[0, t]$ with the uniform metric mentioned in Chapter 4 and $D[0, t]$, the space of all real-valued functions $x(p)$ on $[0, t]$ that are right-continuous and have left limits. A metric which produces a satisfactory set of Borel sets is described in Section 14 of the Billingsley reference.

With the random variables $\{X_n\}$, X and distribution functions $\{F_n\}$, F, there are probability measures P_{X_n} and P_X. Let $\mathcal{A}$ be the set of Borel sets on which $\{P_{X_n}\}$ and P_X are defined. Then $X_n \Rightarrow X$ implies that $P_{X_n}(A) \to P_X(A)$ for all $A \in \mathcal{A}$ such that $P_X(\partial A) = 0$ where the symbol ∂A denotes the boundary of A, which is the set of points which are limits of sequences of points in A and limits of points outside of A. The result that "$X_n \Rightarrow X$ implies that $P_{X_n}(A) \to P_X(A)$ for all $A \in \mathcal{A}$ such that $P_X(\partial A) = 0$" has a converse.

If $\mathcal{A}$ is a set of Borel sets defined explicitly by means of a metric on some space, and the sequence of probability measures $\{P_{X_n}\}$ and P_X are defined on $\mathcal{A}$, then by definition $X_n \Rightarrow X$ if $P_{X_n}(A) \to P_X(A)$ for all $A \in \mathcal{A}$ such that $P_X(\partial A) = 0$. If g is a measurable function, $X_n \Rightarrow X$, and $P_X(D_g) = 0$, then $g(X_n) \Rightarrow g(X)$. This is a very valuable result, and allows us to obtain the limiting distribution of many sequences of random variables.

After these preliminaries, we return to defining risk reserve processes. For the nth process, let $u_n > 0$ be the initial risk reserve, $a_n > 0$ be the gross risk premium per unit time, and $\{x_i^{(n)}, i = 1, 2, ...\}$ be the sequence of independent identically distributed risk sums with $E[X_i^{(n)}] = \mu_n > 0$ and $\sigma^2[X_i^{(n)}] = \sigma_n^2 > 0$. We will specialize Iglehart's process to assume the number of claims follows a Poisson process $\{N(t), t \geq 0\}$ independent of the $X_i^{(n)}$'s, and that with our operational time, the expected time between claims is one unit. The original time scale is compressed by a factor of n^{-1}. Define

$$X_n(t) = u_n + a_n nt - \sum_{i=1}^{N(nt)} X_i^{(n)} \quad \text{for } 0 \leq t \leq 1.$$

The $X_n(\cdot)$ functions are elements of the space $D[0, 1]$.

THEOREM 1. If $u_n = un^{\frac{1}{2}} + o(n^{\frac{1}{2}})$, $a_n = an^{-\frac{1}{2}} + o(n^{-\frac{1}{2}})$,

$$\mu_n = \mu n^{-\frac{1}{2}} + o(n^{-\frac{1}{2}}), \quad \sigma_n \to \sigma^2 > 0,$$

and $E[(X_i^{(n)})^{2+\varepsilon}]$ is bounded in n for some $\varepsilon > 0$, then

$$\lim_{n \to \infty} P\{n^{-\frac{1}{2}} X_n(t) \le x\} = \frac{1}{[2\pi\sigma^2 t]^{\frac{1}{2}}} \int_{-\infty}^{x} \exp\left\{-\frac{(y - [u + (a - \mu)t])^2}{2\sigma^2 t}\right\} dy$$

for all $x \in (-\infty, \infty)$.

The theorem is proved by showing that $X_n \Rightarrow W^*$ for $0 \le t \le T$ where $W^*(t) = u + (a - \mu)t - W(t)$ and $W(t)$ is the Wiener process. The appropriate Wiener distribution is then used.

The nth approximation to the total assets of the company at time t is $X_n(t)$.

We are interested in functionals of $X_n(t)$. One such functional is the time to ruin of the nth process, namely

$$T_n = \inf \{t > 0: X_n(t) \le 0\}$$

if the set $\{t > 0: X_n(t) \le 0\}$ is not empty and $+\infty$ otherwise. Define the random variable

$$T = \inf \{t > 0: u + (a - \mu)t + \sigma W(t) \le 0\} \quad \text{if the set is non-empty}$$

and $+\infty$ otherwise. Here $\{W(t), t \ge 0\}$ is the Wiener process. Note that $P[\text{Time to ruin} > T] = P[Z_T \le u]$ where Z_T was defined in Chapter 3.

THEOREM 2. Under the hypotheses of Theorem 1,

$$\lim_{n \to \infty} P\{T_n \le t\} = \int_0^t f(x) \, dx \quad \text{where for } x > 0,$$

$$f(x) = \frac{c^{-1} e^{-bc}}{(2\pi)^{\frac{1}{2}}} x^{-\frac{3}{2}} \exp\{-\tfrac{1}{2}[c^2 x^{-1} + (bc)^2 x]\},$$

with $b = (a - \mu)/\sigma$ and $c = u/\sigma$.

The proof of Theorem 2 hinges on the fact that the mapping defined by $\tau(x) = \inf \{t > 0: x(t) \le 0\}$ if the set is non-empty and $+\infty$ otherwise is measurable and almost everywhere continuous with respect to the measure corresponding to $u + (a - \mu)t + \sigma W(t), t \ge 0$. Thus $\lim_{n \to \infty} P\{T_n \le t\} = P\{T \le t\}$ and the Wiener measure for the set $\{T \le t\}$ has the value given in Theorem 2.

COROLLARY TO THEOREM 2. Under the hypotheses of Theorem 1, the probability of ultimate ruin for the limit process is given by

$$P\{T<\infty\} = \exp\{-2bc\}.$$

Remark. If we use the definitions of b and c, and the notation from Chapter 3, this says that $\psi(u) \sim \exp(-\lambda u/\mathrm{Var}(X))$, as $u \to \infty$. By comparison, Lundberg has shown that $\psi(u) \sim Ce^{-Ru}$ for appropriate C and R, as $u \to \infty$. In the special case that $P(x) = 1 - e^{-x}$, $x \geq 0$, $C = 1/(1+\lambda)$ and $R = \lambda/(1+\lambda)$. Thus the two approximations are $\exp(-\lambda u)$ and $1/(1+\lambda)\exp(-(\lambda u/(1+\lambda)))$. In this special case, the Lundberg result is an exact result.

Another paper which uses weak convergence of measures and Wiener distributions to approximate collective risk distributions is [5] by Perry M. Gluckman. One of its results concerns the rate of convergence of the approximating processes. This will now be discussed.

For each q, let the sequence $\{X_i^{(q)}, i = 1, 2, \ldots\}$ be independent, identically distributed random variables with $E[X_i^{(q)}] = \mu_q$, $\mathrm{Var}[X_i^{(q)}] = 1$, and satisfying the Lindeberg condition where for each $\varepsilon > 0$

$$E\{[X_i^{(q)}]^2 I\{|X_i^{(q)}| \geq \varepsilon \sqrt{q}\}\} \to 0 \quad \text{as } q \to \infty.$$

The indicator function

$$I\{|X_i^{(q)}| \geq \varepsilon \sqrt{q}\} = \begin{cases} 1 & \text{if } |X_i|^q \geq \varepsilon \sqrt{q} \\ 0 & \text{if } |X_i|^q < \varepsilon \sqrt{q}. \end{cases}$$

Assume that $\alpha > 0, \beta > 0$ and that

$$\alpha_q = \alpha \sqrt{q}, \quad \beta_q = \beta/\sqrt{q}, \quad \text{and} \quad \mu_q = \mu/\sqrt{q}.$$

Then the approximating reserve function is

$$\alpha_q + \beta_q tq - \sum_{i=1}^{N(qt)} X_i^{(q)}, \quad t \geq 0.$$

If the minimum of this function is ≤ 0, ruin has occurred. Let

$$X_q(t) = \sum_{i=1}^{N(qt)} X_i^{(q)} - \beta_q tq, \quad t \geq 0.$$

If
$$\sup_{0 \leq t < \infty} X_q(t) \geq \alpha_q,$$

ruin occurs sometime in the future.

Let $$W^*(t) = W(t) + \mu t - \beta t, \quad t \geq 0.$$

THEOREM 3. If
$$E\{[X_1^{(q)}]^4\} \le R < \infty$$
for all q, and if
$$E\{[X_1^{(q)} - \delta] \mid X_1^{(q)} \ge \delta\} \le R$$
for all $\delta \ge 0$ and q, then
$$\left| P[\sup_{0 \le t < \infty} X_q(t) \ge \alpha \sqrt{q}] - P[\sup_{0 \le t < \infty} W^*(t) \ge \alpha] \right| \le 0\left(\frac{1}{\sqrt{q}}\right).$$

On page 62 of [5], Gluckman states: "The condition that
$$E\{[X_1^{(q)} - \delta] \mid X_1^{(q)} \ge \delta\} \le R < \infty$$
for all q and $\delta \ge 0$ does not place any restrictions on the negative part of the distribution of $X_1^{(q)}$. Two cases where the condition holds are as follows. When
$$dP(X_1^{(q)} \ge x) \sim c_1 e^{-c_2 x}$$
for x large for $x \ge 0$ then for $\delta \ge 0$
$$E\{[X_1^{(q)} - \delta] \mid X_1^{(q)} \ge \delta\} = \frac{1}{c_1 c_2}$$
for δ large. Also if
$$dP(X_1^{(q)} \ge x) = e^{-x^2/2}/\sqrt{2\pi}$$
for $x \ge 0$, we have
$$E\{[X_1^{(q)} - \delta] \mid X_1^{(q)} \ge \delta\} \le E\{X_1^{(q)} \mid X_1^{(q)} \ge 0\}."$$

Note that
$$P[\sup_{0 \le t < \infty} W^*(t) \le \alpha] = P[\sup_{0 \le t < \infty} [W(t) - (\beta - \mu)t] \ge \alpha]$$
$$= P[\sup_{0 \le t < \infty} [W(t) - \alpha - (\beta - \mu)t] \ge 0]$$
$$= P[T < \infty]$$
where
$$T = \inf\{t \ge 0 : \alpha + (\beta - \mu)t - W(t) \le 0\}.$$

Now $P[T<\infty]=P[T^*<\infty]$ by the symmetry of the Wiener process where

$$T^* = \inf\{t\geq 0: \alpha+(\beta-\mu)t+W(t)\leq 0\}.$$

The $P[T^*<\infty]$ has been recorded in the Corollary to Theorem 2.

Gluckman points out that these approximations should be interpreted "... in the sense that if the initial reserve α is large and the mean cash flow $(\beta-\mu)>0$ is small, then the reserves

$$\left(\beta t - \sum_{i=1}^{N(t)} X_i\right)$$

will behave like a Wiener process with drift."

(See [5], page 15.) The reader will observe that $\alpha_q \to +\infty$ and $(\beta_q-\mu_q)\to 0$ as $q\to\infty$.

In reference [4] Hans Gerber uses a linear combination of a compound Poisson and a diffusion process to analyze certain collective risk problems.

Let the income process $X(t)$ of an insurance firm be given by $X(t)=u+(p_1+\lambda)t-X_1(t)+X_2(t)$, $t\geq 0$, where $X_1(t)$ is a compound Poisson process, and $X_2(t)$ is a diffusion process representing an additional (but independent) uncertainty of the income process. We have

$$P[X_2(t)\leq x] = \frac{1}{\sqrt{4\pi Dt}}\int_{-\infty}^{x} \exp\left(\frac{-y^2}{4Dt}\right) dy.$$

The collective risk theory in Chapter 3 had $D=0$, i.e. $X_2(t)=0$.

By using an asymptotic result for the solution of the extended defective renewal equation developed in the paper, Gerber proves the following:

THEOREM. $\psi(u) \sim C e^{-ku}$ for $u\to\infty$, where

$$C = \lambda \bigg/ \left[k\int_0^\infty y e^{ky}(1-F(y))\, dy + kD\right]$$

provided that the equation

$$1 - \frac{D}{p_1+\lambda} k = \frac{1}{p_1+\lambda}\int_0^\infty e^{ky}[1-F(y)]\, dy$$

has a solution k.

When $D=0$, this reduces to Lundberg's asymptotic approximation for $\psi(u)$.

One may consider the collective risk process as random walk in one dimension, as H. Bohman does in [1]. A step occurs at the time of a claim, and has length equal to the amount of the claim minus the amount of premium collected since the preceding claim. A positive difference implies a step to the right; a negative difference produces a step to the left. There is a barrier u units to the right of the origin. It is now seen that the probability of ruin is equivalent to the first passage probability through u. Making use of an assumption that u is large, and that λ is small compared with u, Bohman derives the first passage probability in the form

$$\psi(u, T) = \int_0^{T\sigma^2/u^2} \frac{1}{\sqrt{2\pi}} z^{-\frac{3}{2}} \exp\left\{-\frac{[1 - \lambda u z/\sigma^2]^2}{2z}\right\} dz.$$

This is essentially the same form as given in Theorem 2. Our use of operational time in Iglehart's results causes some difficulty in comparing the constants in the density. Jan Grandell cautions the reader to only use the Wiener approximation for u large, λ very small by showing that Bohman's assumptions are not only sufficient, but also necessary [6].

Consider the class of risk processes characterized by the equation

$$F(x, t) = \int_0^{x\sqrt{b}} \frac{1}{\Gamma(bt)} y^{bt-1} e^{-y} dy \quad \text{for } b > 0.$$

Harald Bohman studies these processes in [2], and observes that when $b \to \infty$, the limiting process is the Wiener process. He derives the result that

$$1 - \psi(1, T) = \frac{1}{\sqrt{2\pi}} \int_0^{T^{-\frac{1}{2}}} e^{-x^2/2} dx - \int_0^T \frac{e^{-1/(2x)} dx}{2\pi \sqrt{x(T-x)}}.$$

The restriction that $u = 1$ is lessened by choosing the initial value of the risk reserve as the monetary unit. It is assumed that $\lambda = 0$. An expression which is easier to calculate is given and tabulated for various T's.

REFERENCES

1. BOHMAN, H. (1972). Risk theory and Wiener processes. *Astin Bulletin* **7**, 96–99.
2. — (1972). A class of risk processes where explicit formulas for the ruin-probabilities can be obtained. *Skand. Aktuarietidskr.* **55**, 25–27.
3. DOOB, J. L. (1953). *Stochastic Processes*. Wiley, New York.
4. GERBER, HANS U. (1970). An extension of the renewal equation and its application in the collective theory of risk. *Skand. Aktuarietidskr.* **53**, 205–210.
5. GLUCKMAN, P. M. (1970). Applications of diffusion approximations to the collective theory of risk. Tech. Report No. 23, Dept. of Statistics, Stanford University.
6. GRANDELL, J. (1972). A remark on Wiener process approximation of risk processes. *Astin Bulletin*, **7**, 100–101.
7. IGLEHART, D. L. (1969). Diffusion approximations in collective risk theory. *J. Applied Prob.* **6**, 285–292.

CHAPTER 6

APPLICATIONS IN STATISTICS

6.1. Kolmogorov statistics

One of the most appealing concepts from undergraduate statistics is that of the empirical distribution function. If $X_1, X_2, \ldots, X_n$ is a random sample from a population with (unknown) distribution function $F(x)$, then its empirical distribution function is defined by

$$F_n(x) = \frac{1}{n}(\# \text{ of } X_i\text{'s} \leq x)$$

$$= \frac{1}{n}\sum_{i=1}^{n}\psi_x(X_i), \quad -\infty < x < \infty,$$

where

$$\psi_y(x) = \begin{cases} 1 & \text{if } x \leq y \\ 0 & \text{if } x > y \end{cases}.$$

For each x, $F_n(x)$ is a random variable. Hence $\{F_n(x), -\infty < x < \infty\}$ is a stochastic process. Now

$$E\{F_n(x)\} = \frac{1}{n}\sum_{i=1}^{n} E\{\psi_x(X_i)\}$$

$$= F(x)$$

since $E\{\psi_x(X_i)\} = 1 F(x) + 0[1 - F(x)]$. Thus the mean function is as we would want it to be. Furthermore, it is easy to show that

$$\text{Var}\{F_n(x)\} = \frac{1}{n} F(x)[1 - F(x)]$$

which reveals that as $n \to \infty$, $F_n(x)$ becomes concentrated around $F(x)$. In fact, by the Glivenko–Cantelli Lemma (see, e.g. Loève [12]),

$$P[\lim_{n\to\infty} \sup_{-\infty < x < \infty} |F_n(x) - F(x)| = 0] = 1.$$

By the Central Limit Theorem,

$$\lim_{n\to\infty} P\left[\frac{\sum_{i=1}^{n}\psi_x(X_i)-nF(x)}{\sqrt{nF(x)\left[1-F(x)\right]}}\leqslant y\right]$$

$$=\lim_{n\to\infty} P\left[\frac{F_n(x)-F(x)}{\{F(x)\left[1-F(x)\right]/n\}^{\frac{1}{2}}}\leqslant y\right]$$

$$=\frac{1}{\sqrt{2\pi}}\int_{-\infty}^{y} e^{-w^2/2}dw.$$

Before we go further, let us compare the development so far with a simple statistical problem.

Let $X_1, X_2, ..., X_n$ be a random sample from a population determined by

$$X=\begin{cases} 1 & \text{with (unknown) probability } p \\ 0 & \text{with probability } 1-p \end{cases}$$

If $\bar{X}_n = (X_1+X_2+...+X_n)/n$, then $E(\bar{X}_n)=p$, $\text{Var}(\bar{X}_n)=p(1-p)/n$,

and
$$\lim_{n\to\infty} P\left[\frac{\bar{X}_n - p}{[p(1-p)/n]^{\frac{1}{2}}}\leqslant y\right]$$

$$=\frac{1}{\sqrt{2\pi}}\int_{-\infty}^{y} e^{-w^2/2}dw.$$

Just as $\bar{X}_n$ is an estimate for p, $F_n(y)$ is an estimate of $F(y)$. Of course to be of any value, we need a simultaneous estimate of $F(y)$, $-\infty < y < \infty$, and probabilities relevant to the estimator. For this purpose, we use the following statistics.

DEFINITIONS. $D_n = \text{l.u.b.}_{-\infty < y < \infty} |F_n(y) - F(y)|$ is a random variable called the two-sided Kolmogorov statistic. The one-sided Kolmogorov statistics are

$$D_n^+ = \underset{-\infty<y<\infty}{\text{l.u.b.}} [F_n(y) - F(y)],$$

and
$$D_n^- = \underset{-\infty<y<\infty}{\text{l.u.b.}} [F(y) - F_n(y)].$$

(See [9] e.g. and its references to the Kolmogorov papers; also [4] and [7].)

The distributions of D_n, D_n^+, and D_n^-, for finite n and $n\to\infty$, have been tabulated, and with these distributions, one can:

(1) construct confidence bands for the unknown $F(x)$;
(2) test a hypothesis about $F(x)$.

The ability to calculate these distributions relies heavily on the following theorem.

THEOREM 1. (Kolmogorov.) For any positive integer n, the probability distribution of D_n is the same for any continuous distribution $F(x)$, $-\infty < x < \infty$, and, in particular, is the same as for the uniform distribution on $[0, 1]$.

Proof. The transformation

$$H[X] = \int_{-\infty}^{X} dF(x) = Y$$

ascribes to each sample $X_1, X_2, ..., X_n$ of X a sample $Y_1, Y_2, ..., Y_n$ of Y (from the uniform distribution) and to $F_n(x)$, $-\infty < x < \infty$ another empirical distribution function $G_n(y)$, $0 \leq y \leq 1$, while the distribution function $F(x)$, $-\infty < x < \infty$ is transformed into the distribution function $G(y) = y$ for $0 \leq y \leq 1$. For any sample $X_1, X_2, ..., X_n$ and the corresponding sample $Y_1, Y_2, ..., Y_n$ we have

$$D_n = \underset{-\infty < x < \infty}{\text{l.u.b.}} |F_n(x) - F(x)|$$

$$= \underset{0 \leq y \leq 1}{\text{l.u.b.}} |G_n(y) - y|$$

$$= D'_n.$$

Hence, for any real numbers $a < b$, $P[a < D_n < b] = P[a < D'_n < b]$.

Kolmogorov derived the limiting distribution of D_n.

THEOREM 2. (Kolmogorov). If $F(x)$ is continuous, then for any $z > 0$,

$$\lim_{n \to \infty} P[D_n \leq z/n^{\frac{1}{2}}] = 1 - 2 \sum_{j=1}^{\infty} (-1)^{j-1} \exp\{-2j^2 z^2\}.$$

It is possible to give a proof of Theorem 2 using the results of weak convergence from Chapter 5. Let a sequence of random functions Z_n be defined by

$$Z_n(t) = \sqrt{n} |F_n(t) - F(t)|,$$

or

$$Z_n(t) = \sqrt{n} |F_n(t) - t|$$

for $0 \leqslant t \leqslant 1$ since we may assume that $F(t)$ is uniformly distributed on the interval $[0, 1]$. The function Z_n is a random function taking values in $D[0, 1]$ for which an approprate metric can be specified. Now $Z_n \Rightarrow Z$ where $Z(t)$ is normal with $E(Z(t)) = 0$, $E[Z(s)Z(t)] = s(1-t)$ for $0 \leqslant s \leqslant t \leqslant 1$, i.e. $Z(t)$, $0 \leqslant t \leqslant 1$ is the Doob-Kac process from Chapter 4. Note that for almost all sample points, $Z_n(0) = 0$, for all sample points $Z_n(1) = 0$, and that $Z(0) = Z(1) = 0$. The function $h(Z) = \sup_{0 \leqslant t \leqslant 1} |Z(t)|$ has $P_Z(D_h) = 0$ for the appropriate metric, so that $h(Z_n) \Rightarrow h(Z)$. By using distribution K from section 4.7, we have the asymptotic distribution of the Kolmogorov statistic $h(Z_N)$.

THEOREM 3. If $F(x)$ is continuous, then for any $z > 0$,
$$\lim_{n \to \infty} P[D_n^+ \leqslant z/n^{\frac{1}{2}}]$$
$$= \lim_{n \to \infty} P[D_n^- \leqslant z/n^{\frac{1}{2}}] = 1 - e^{-2z^2}.$$

The method of proof is that of Theorem 2. For D_n^+, one uses $h(Z) = \max_{0 \leqslant t \leqslant 1} Z(t)$. For D_n^-, one uses $h(Z) = -\min_{0 \leqslant t \leqslant 1} Z(t)$. The equality of limiting distributions makes use of the fact that the $-Z(t)$ process is stochastically identical with the $Z(t)$ process. The proof uses distribution J from section 4.7.

The reader may also wish to consult reference [7] by Doob in connection with Theorems 2 and 3.

The distributions of D_n, D_n^+, and D_n^- for finite n, can be found in the references given in the next section. They can be used to build confidence bands for an unknown distribution function $F(x)$. We illustrate this with D_n.

THEOREM 4. If $F(x)$ is continuous, and D_n^α is a number such that $P[D_n \leqslant D_n^\alpha] = 1 - \alpha$, then $F_n(x) - D_n^\alpha$ and $F_n(x) + D_n^\alpha$ form a confidence band with confidence coefficient $1-\alpha$ for $F(x)$.

Proof. $1 - \alpha = P\{\operatorname*{l.u.b.}_{-\infty < x < \infty} |F_n(x) - F(x)| \leqslant D_n^\alpha\}$
$= P\{|F(x) - F_n(x)| \leqslant D_n^\alpha \text{ for all } x\}$
$= P\{F_n(x) - D_n^\alpha \leqslant F(x) \leqslant F_n(x) + D_n^\alpha \text{ for all } x\}.$

6.2. KAC STATISTICS

All of section 6.1 assumes a *fixed sample size*. This may not be appropriate in some situations in biology, insurance, and telephone engineering where you wish to observe random variables over a *fixed time period*. For example, if you wanted to study the distribution of insurance claims during the next six months the number of claims would be a random variable usually assumed to have a Poisson distribution. The same is true if you examine telephone calls for one week, or the number of insects trapped in three hours. There are reasons for ignoring the randomness of the sample size, which will be mentioned later, but if you wish your analysis to allow for it, the following method is appropriate.

Let $N, X_1, X_2, \ldots$ be independent random variables, N having a Poisson distribution with mean λ and each X_i having the same continuous distribution function $F(y)$. Let $\psi_y(x)$ be 0 or 1 according as $x>y$ or $x \leqslant y$. A modified empirical distribution function was defined by Mark Kac in 1949 (see [11]) as

$$F_\lambda^*(y) = \frac{1}{\lambda} \sum_{j=1}^{N} \psi_y(X_j), \quad -\infty < y < \infty,$$

where the sum is taken to be zero if $N=0$. Notice that it is possible for $F_\lambda*(y)$ to exceed one. The statistic analogous to D_n^- is

$$K_\lambda^- = \operatorname*{l.u.b.}_{-\infty < y < \infty} [F(y) - F_\lambda^*(y)].$$

We call this the one-sided Kac statistic. The fact that $F_\lambda*(y)$ may exceed one is not as serious a draw-back as it may appear. We will be interested in an upper confidence contour for $F(y)$ of the form:

$$F_{\lambda, \varepsilon}^*(y) = \min \, [F_\lambda^*(y) + \varepsilon, 1], \quad -\infty < y < \infty.$$

We will need to know $P\{F(y) \leqslant F_{\lambda, \varepsilon}^*(y), -\infty < y < \infty\}$ but it is easy to see that that is the same as

$$P\{\operatorname*{l.u.b.}_{-\infty < y < \infty} [F(y) - F_\lambda^*(y)] \leqslant \varepsilon\},$$

which is computed in the next theorem.

THEOREM 1. *For $N, X_1, X_2, \ldots$ as before and $0 < \varepsilon \leqslant 1$,*

$$P[K_\lambda^- \leqslant \varepsilon] = 1 - \varepsilon\lambda \sum_{j=0}^{[\lambda(1-\varepsilon)]} \frac{(\lambda\varepsilon + j)^{j-1}}{j!} e^{-\lambda\varepsilon - j}$$

where the upper limit of summation involves the greatest integer function.

Sketch of proof (see [1]). Kac notes, as Kolmogorov also did, that as long as $F(y)$ is continuous, the distribution of the statistic K_λ^- is independent of $F(y)$. So you may and we do assume $F(y) = y$, $0 \leqslant y \leqslant 1$.

By the independence of N, X_1, X_2, ..., the distribution free property, and the distribution of N,

$$P[K_\lambda^- \leq \varepsilon] = \sum_{n=0}^{\infty} \frac{\lambda^n e^{-\lambda}}{n!} P\left\{y \leq \min\left[1, \frac{1}{\lambda} \sum_{i=1}^{n} \psi_y(X_i) + \varepsilon\right], \ 0 \leq y \leq 1\right\}.$$

The inner probability may be found by slightly changing the method contained in a 1951 *Annals* paper by Birnbaum and Tingey [4]. After substituting the resulting expression, and interchanging order of summation and summing on n, we get the desired result.

The asymptotic distribution of the one-sided Kolmogorov statistic D_n^- has been found by Smirnov to be

$$\lim_{n \to \infty} P\left\{\underset{-\infty < y < \infty}{\text{l.u.b.}} [F(y) - F_n(y)] \leq \frac{\alpha}{\sqrt{n}}\right\} = 1 - e^{-2\alpha^2}, \quad 0 \leq \alpha < \infty.$$

The analogous theorem for the one-sided Kac statistic is:

THEOREM 2. *For N, X_1, X_2, ... subject to the previous conditions,*

$$\lim_{\lambda \to \infty} P\left\{\underset{-\infty < y < \infty}{\text{l.u.b.}} [F(y) - F_\lambda^*(y)] \leq \frac{\alpha}{\sqrt{\lambda}}\right\}$$

$$= [1/\pi]^{\frac{1}{2}} \int_0^\alpha e^{-u^2/2} du, \quad \alpha \geq 0$$

$$= 0, \quad \alpha < 0.$$

Sketch of proof (similar to analysis of M. Kac for the asymptotic distribution of the two-sided statistic). Since the probability on the left side above is independent of $F(y)$ it suffices to let $F(y) = y$, $0 \leq y \leq 1$. Consider the process

$$x_\lambda(y) = \lambda^{\frac{1}{2}}\left\{y - \frac{1}{\lambda} \sum_{j=1}^{N} \psi_y(X_j)\right\}, \quad 0 \leq y \leq 1,$$

with independent increments. Without loss of generality, we will consider the separable version of the process. Using the distribution—free property, separability, and the monotonicity of the sets, we obtain

$$\lim_{\lambda \to \infty} P\left\{\underset{-\infty < y < \infty}{\text{l.u.b.}} [F(y) - F_\lambda^*(y)] \leq \frac{\alpha}{\sqrt{\lambda}}\right\}$$

$$= \lim_{\lambda \to \infty} P\left\{\lim_{r \to \infty} \underset{1 \leq k \leq 2^r}{\text{l.u.b.}} x_\lambda\left(\frac{k}{2^r}\right) \leq \alpha\right\}$$

$$= \lim_{\lambda \to \infty} \lim_{r \to \infty} P\left\{\underset{1 \leq k \leq 2^r}{\text{l.u.b.}} x_\lambda\left(\frac{k}{2^r}\right) \leq \alpha\right\}.$$

Using the fact that the $x_\lambda(y)$ process has independent increments, by similar reasoning to that of Erdös and Kac, [8] the above equals

$$P\{\underset{0\leqslant u\leqslant 1}{\text{l.u.b.}}\, x(u)\leqslant \alpha\}$$

where $\{x(u),\ 0\leqslant u\leqslant 1\}$ is the Wiener process. The conclusion follows from distribution A in Section 4.7.

Remark. Thus for large λ, we can evaluate $P_\lambda(\varepsilon) \doteq 2F(\alpha)-1$ where $\varepsilon=\alpha/\sqrt{\lambda}$ and $F(\alpha)$ is the area to the left of $\alpha(\alpha\geqslant 0)$ under a standardized normal curve. Our tables show that even for $\lambda=25$, the true and asymptotic values are close together. In many practical situations, λ would exceed 50 for an appropriate observation period.

Power and consistncy of the one-sided test

Assume we are testing the hypothesis that $F(x)=\theta(x)$ against the alternative $F(x)=\Phi(x)$ where $\sup_{-\infty<x<\infty}[\theta(x)-\Phi(x)]=\delta>0$.

Assume that for fixed λ, and a test of size α, we pick ε_λ corresponding to $1-\alpha$. Modifying the analysis of Birnbaum, [3], it is possible to find a lower bound on the power of the test (probability that we reject when the sample has $\Phi(x)$ for distribution).

$$\text{Power} = P[F_\lambda^*(x)+\varepsilon_\lambda < \theta(x) \quad \text{for some } x;\ \Phi(x)]$$
$$\geqslant \sum_{i=0}^{k} e^{-\lambda u_0}(\lambda u_0)^i/i!$$

where $u_0=\theta(x_0)-\delta$, x_0 is determined by $\theta(x_0)-\Phi(x_0)=\delta$ and $k=$greatest integer in $\lambda\{\theta(x_0)-\varepsilon_\lambda\}$.

If we use the phrase "modified consistent" to indicate that the limiting value of the power as $\lambda\to\infty$ is one, one can show that the one-sided Kac statistic is modified consistent, by using the above lower bound on the power.

THEOREM. For N, X_1, X_2, ..., $\theta(x)$, $\Phi(x)$, δ, α, and ε_λ subject to the previous conditions, $\lim_{\lambda\to\infty}\{F_\lambda^*(x)+\varepsilon_\lambda<\theta(x)\text{ for some } x;\ \Phi(x)\}=1$

Proof. We shall assume that $u_0=\theta(x_0)-\delta=\Phi(x_0)>0$. Let $\beta>0$. For fixed α, with $P\{\text{l.u.b.}_{-\infty<y<\infty}\ [F(y)-F_\lambda^*(y)]\leqslant\varepsilon_\lambda\}=1-\alpha$, we see that $\lim_{\lambda\to\infty}\varepsilon_\lambda=0$. Hence one can obtain Λ_1, and a positive constant c independent of λ such that $\lambda>\Lambda_1$ implies that $[\lambda(\theta(x_0)-\varepsilon_\lambda)]/(\lambda u_0)\geqslant 1+c$. Using this fact and the lower bound on the power, for $\lambda>\Lambda_1$,

$$P\{F_\lambda^*(x)+\varepsilon_\lambda<\theta(x)\text{ for some } x;\ \Phi(x)\}\geqslant P\{X\leqslant(1+c)\lambda u_0\}$$

where X has a Poisson distribution with mean λu_0. There exists Λ_2 such that

$$I = \int_{-\infty}^{c\sqrt{\Lambda_2 u_0}} \frac{e^{-x^2/2}}{\sqrt{2\pi}} dx \geq 1 - \frac{\beta}{2}.$$

By the limiting distribution for X, (see, for example, p. 245 of reference [13] in Chapter 3), there exists Λ_3 such that $\lambda > \max(\Lambda_2, \Lambda_3)$ implies that $P\{X \leq (1+c)\lambda u_0\} \geq P\{(X-\lambda u_0)/\sqrt{\lambda u_0} \leq c\sqrt{\Lambda_2 u_0}\} \geq I - \beta/2$. Hence for $\lambda > \max(\Lambda_1, \Lambda_2, \Lambda_3)$, $P\{F_\lambda^*(x) + \varepsilon_\lambda < \theta(x)$ for some x; $\Phi(x)\} \geq 1 - \beta$. Since β was arbitrary, the theorem is proved.

Remark. The following drawing may serve to illustrate the conditional probability in the theorem. It is exaggerated to show the point.

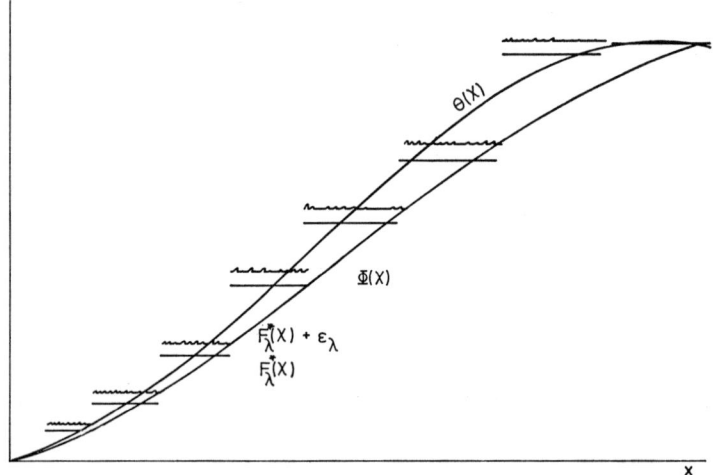

To repeat an earlier remark, the tables which follow can be used to
(1) construct confidence bands for an unknown $F(x)$
(2) test a hypothesis about $F(x)$.

The finite λ table also serves for lower confidence contours since

$$P\{ \underset{-\infty < y < \infty}{\text{g.l.b.}} [F(y) - F_\lambda^*(y)] \geq -\varepsilon \}$$
$$= P\{ \underset{-\infty < y < \infty}{\text{l.u.b.}} [F(y) - F_\lambda^*(y)] \leq \varepsilon \}, \quad 0 < \varepsilon \leq 1.$$

Two-sided statistic (see [2])

Consider the two-sided Kac statistic

$$\underset{-\infty < y < \infty}{\text{l.u.b.}} |F(y) - F_\lambda^*(y)| = K_\lambda.$$

The finite λ distribution was solved from a system of difference equations which made use of a 1950 *Annals* paper by Massey, [13].

The asymptotic distribution was derived by Kac and turns out to be the distribution for the absolute maximum functional of the Wiener process

$$P\{\max_{0\leqslant u\leqslant 1}|x(u)|\leqslant \alpha\}=\frac{4}{\pi}\sum_{k=0}^{\infty}\frac{(-1)^k}{2k+1}\exp\left\{-\frac{(2k+1)^2\pi^2}{8\alpha^2}\right\},\ \alpha>0.$$

Recall that for the Doob-Kac process (See [7])

$$P\{\max_{0\leqslant u\leqslant 1}|x(u)|\leqslant \alpha\}=1-2\sum_{k=1}^{\infty}(-1)^{k+1}e^{-2k^2\alpha^2}.$$

An alternate expression is given in a 1948 *Annals* paper by W. Feller [9] which is more analogous. For $\alpha>0$,

$$P\{\max_{0\leqslant u\leqslant 1}|x(u)|\leqslant \alpha\}=\frac{\sqrt{2\pi}}{\alpha}\sum_{k=0}^{\infty}\exp\left\{-\frac{(2k+1)^2\pi^2}{8\alpha^2}\right\}.$$

It is also possible to obtain a lower bound on the power of the test based on the two-sided Kac statistic and show that it approaches 1 as $\lambda\to\infty$.

Criticisms. The confidence bands for the Kac statistics are wider than for the Kolmogorov statistics with $\lambda=n$. One therefore may prefer to ignore the fact that the number of observations in a fixed time period is a Poisson variate (in some cases) and consider a fixed sample size instead.

The criticism that F_λ^* can exceed one can be obviated by using a statistic invented by L. Takács [15]. It is based on a more appealing modified empirical distribution function:

$$\frac{1}{N}\sum_{j=1}^{N}\psi_y(X_j).$$

His paper derives the exact distribution for the allied distribution, but does not tabulate it. Probably the confidence bands would still be wider then the analogous Kolmogorov confidence bands.

Despite these criticisms, for the *fixed time periods* used in insurance, biology, and telephone engineering, the Kac statistic may be useful to test hypotheses and to build confidence bands.

Further references. The weak convergence of the empirical process with random sample size was studied by R. Pyke in [14], P. Fernandez in [10], and M. Csörgö and S. Csörgö in [5]. M. Csörgö and M. Alvo derived explicit expressions for

$$P\{\operatorname*{l.u.b.}_{-\infty<y\leqslant y_b} [F(y) - F_\lambda^*(y)] \leqslant \varepsilon\},$$

and

$$P\left\{\operatorname*{l.u.b.}_{-\infty<y\leqslant y_b} \frac{F(y) - F_\lambda^*(y)}{1 - F(y)} \leqslant \varepsilon\right\}$$

where y_b is a real number with $F(y_b) = b$, and such that b is somewhat restricted. (See [6].)

Exercises

1. An insurance company decides to approximate the distribution of claims. It decides to observe for six months. Past records indicate that the average number of claims for six months is 35, i.e. $\lambda = 35$. The sample produced 37 claims which (in thousands of dollars) were: 1.1, 1.1, 1.6, 1.7, 1.8, 1.8, 1.9, 1.9, 2.0, 2.1, 2.2, 2.2, 2.3, 2.4, 2.4, 2.5, 2.5, 2.5, 2.6, 2.6, 2.6, 2.7, 2.9, 3.0, 3.0, 3.0, 3.5, 3.5, 3.5, 4.0, 4.0, 4.0, 4.0, 4.5, 4.6, 4.7, 4.8.
Build an 85% upper confidence contour.

2. Since the sample mean $\bar{x} = 2.8$, and the sample variance $s^2 = .97$, the insurance company decides to test a null hypothesis that the distribution of claims is normal with mean 2.5 and variance 1. Would it reject this hypothesis at the 15% level of significance?

3. If the Poisson variate's mean had been larger, namely 50, what approximate $\varepsilon/\sqrt{\lambda}$ would correspond to $\lambda = 50$ for 90% probability for both the one-sided and the two-sided Kac statistic? Do the same for $\lambda = 100$ and 95% probability.

Table 1. $P\{\text{l.u.b.}_{-\infty < y < \infty}[F(y) - F_\lambda^*(y)] \leqslant \varepsilon\}^a$

ε	λ						
	5	10	15	20	25	30	35
.001	.00187	.00258	.00314	.00361	.00402	.00440	.00475
.01	.01859	.02571	.03125	.03595	.04010	.04386	.04732
.025	.04621	.06388	.07763	.08926	.09953	.10881	.11735
.05	.09146	.12629	.15325	.17597	.19982	.21739	.23353
.075	.13564	.18692	.23386	.26555	.29345	.32366	.34601
.1	.17868	.24551	.30583	.34623	.38859	.41901	.45241
.125	.22051	.31669	.37402	.43159	.47977	.51355	.55066
.15	.26106	.37256	.45177	.50270	.55574	.60056	.63914
.175	.30029	.42553	.51263	.57973	.63390	.67883	.71677
.2	.33817	.47549	.56876	.63892	.69423	.73903	.77598
.225	.41218	.54587	.63697	.70454	.75677	.79816	.83151
.25	.44931	.59032	.68364	.75090	.81109	.84775	.87658
.275	.48465	.63142	.74303	.80348	.84804	.88820	.91209
.3	.51823	.66928	.77969	.83753	.88681	.91450	.93921
.325	.55009	.73045	.81194	.87702	.91814	.94008	.95927
.35	.58025	.76137	.85553	.90035	.93672	.95931	.97360
.375	.60876	.78927	.87877	.92802	.95644	.97330	.98348
.4	.63568	.81436	.89869	.94287	.96714	.98085	.98873
.425	.71636	.86046	.92720	.96086	.97857	.98812	.99335
.45	.73916	.87889	.94024	.96958	.98659	.99292	.99623
.475	.76032	.89512	.95976	.98038	.99030	.99596	.99795
.5	.77993	.90936	.96758	.98508	.99431	.99729	.99894
.55	.81489	.94805	.98407	.99331	.99780	.99927	.99975
.6	.84473	.96265	.99015	.99729	.99924	.99978	.99994
.65	.91487	.98196	.99588	.99902	.99984	.99996	.99999
.7	.93092	.98756	.99851	.99969	.99996	.99999	1.00000
.75	.94404	.99525	.99956	.99992	.99999	1.00000	1.00000
.8	.95473	.99686	.99976	.99998	1.00000	1.00000	1.00000
.85	.98574	.99916	.99995	1.00000	1.00000	1.00000	1.00000
.9	.98889	.99947	.99999	1.00000	1.00000	1.00000	1.00000
.95	.99135	.99993	1.00000	1.00000	1.00000	1.00000	1.00000
.99	.99292	.99995	1.00000	1.00000	1.00000	1.00000	1.00000

[a] These tables are reprinted with the permission of the editors of *The Annals of Mathematical Statistics*.

Table 2. $P\{\text{l.u.b.}_{-\infty<y<\infty}[F(y)-F_\lambda^*(y)] \leqslant \varepsilon/\sqrt{\lambda}\}$

	λ					λ			
ε	25	30	35	$\lim_{\lambda\to\infty}$	ε	25	30	35	$\lim_{\lambda\to\infty}$
.025	.02009	.02006	.02004	.0200	.825	.60751	.60242	.59900	.5906
.05	.04010	.04005	.04002	.0398	.85	.62087	.61582	.61958	.6046
.075	.06002	.05995	.05991	.0598	.875	.63390	.62891	.63273	.6184
.1	.07983	.07976	.07971	.0796	.9	.64661	.64168	.64556	.6318
.125	.09953	.09945	.09941	.0996	.925	.65899	.66265	.65807	.6450
.15	.11910	.11903	.11899	.1192	.95	.67105	.67478	.67025	.6578
.175	.13854	.13847	.14042	.1390	.975	.68280	.68658	.68212	.6706
.2	.15783	.16041	.16000	.1586	1.00	.69423	.69807	.69368	.6827
.225	.18050	.17987	.17943	.1782	1.1	.74696	.74938	.74398	.7286
.25	.19982	.19915	.19869	.1974	1.2	.78442	.78699	.78906	.7698
.275	.21895	.21825	.21778	.2168	1.3	.82667	.82781	.82212	.8064
.3	.23789	.23717	.23669	.2358	1.4	.85466	.85595	.85710	.8384
.325	.25662	.25589	.25540	.2548	1.5	.88681	.88686	.88136	.8664
.35	.27515	.27441	.27776	.2736	1.6	.90663	.90685	.90723	.8904
.375	.29345	.29749	.29628	.2924	1.7	.92979	.92908	.92878	.9108
.4	.31153	.31582	.31458	.3108	1.8	.94304	.94256	.94241	.9282
.425	.33576	.33391	.33265	.3292	1.9	.95881	.95772	.95711	.9426
.45	.35363	.35175	.35047	.3472	2.0	.96714	.96632	.96588	.9544
.475	.37125	.36934	.36805	.3652	2.1	.97724	.97610	.97541	.9642
.5	.38859	.38667	.38538	.3830	2.2	.98215	.98350	.98269	.9722
.525	.40567	.40374	.40784	.4004	2.3	.98821	.98723	.98660	.9786
.55	.42247	.42710	.42480	.4176	2.4	.99092	.99155	.99091	.9836
.575	.43898	.44380	.44148	.4348	2.5	.99431	.99357	.99308	.9876
.6	.45522	.46020	.45788	.4514	2.6	.99569	.99594	.99549	.9906
.625	.47977	.47631	.47399	.4682	2.7	.99745	.99696	.99662	.9930
.65	.49558	.49213	.48980	.4844	2.8	.99811	.99818	.99789	.9948
.675	.51109	.50764	.50532	.5004	2.9	.99895	.99866	.99872	.9962
.7	.52628	.52285	.52708	.5160	3.0	.99924	.99924	.99907	.9974
.725	.54117	.53775	.54209	.5316	3.1	.99961	.99945	.99946	.9980
.75	.55574	.56028	.55679	.5468	3.2	.99972	.99971	.99962	.9986
.775	.57000	.57465	.57117	.5618	3.3	.99987	.99985	.99979	.9990
.8	.58394	.58869	.58524	.5762	3.4	.99991	.99990	.99989	.9994

Table 3. $P\{\text{l.u.b.}_{-\infty<y<\infty} | F(y) - F_\lambda^*(y)| < k/\lambda\}$

k	λ=1	2	3	4	5	6
1	.36788	.13534	.04979	.01832	.00674	.00248
2		.69923	.53106	.40447	.30845	.23534
3			.84887	.75285	.66479	.58646
4				.91866	.86655	.81194
5					.95350	.92542
6						.97218

k	λ=7	8	9	10	15	20
1	.00091	.00034	.00012	.00005	.00000	.00000
2	.17960	.13708	.10462	.07985	.02068	.00536
3	.51733	.45642	.40276	.35545	.19048	.10213
4	.75818	.70676	.65825	.61282	.42805	.29909
5	.89313	.85873	.82360	.78861	.62806	.49792
6	.95680	.93805	.91684	.89398	.77250	.65893
7	.98281	.97415	.96324	.95038	.86857	.77841
8		.98915	.98410	.97768	.92798	.86205
9			.99305	.99002	.96224	.91756
10				.99550	.98081	.95253
11					.99042	.97351
12					.99525	.98557
13					.99765	.99227
14					.99886	.99590
15					.99945	.99784
16						.99888
17						.99942
18						.99971
19						.99986
20						.99993

Table 3 (cont.)

k	λ					
	25	30	35	40	45	50
1	.00000	.00000	.00000	.00000	.00000	.00000
2	.00139	.00036	.00009	.00002	.00001	.00000
3	.05476	.02936	.01574	.00844	.00453	.00243
4	.20905	.14613	.10215	.07141	.04992	.03490
5	.39462	.31278	.24793	.19654	.15581	.12352
6	.56032	.47614	.40456	.34375	.29209	.24820
7	.69269	.61485	.54527	.48343	.42857	.37993
8	.79258	.72525	.66220	.60403	.55072	.50202
9	.86470	.80957	.75528	.70328	.65417	.60816
10	.91460	.87170	.82701	.78241	.73895	.69719
11	.94775	.91592	.88064	.48382	.80668	.77000
12	.96891	.94633	.91958	.89025	.85953	.82827
13	.98194	.96657	.94704	.92446	.89983	.87395
14	.98970	.97963	.96587	.94905	.92988	.90904
15	.99422	.98782	.97844	.96629	.95179	.93544
16	.99679	.99283	.98662	.97809	.96742	.95492
17	.99823	.99583	.99182	.98599	.97834	.96901
18	.99903	.99760	.99507	.99118	.98582	.97901
19	.99948	.99863	.99706	.99451	.99084	.98598
20	.99972	.99923	.99826	.99663	.99416	.99076
21	.99985	.99957	.99891	.99795	.99632	.99398
22	.99992	.99976	.99941	.99876	.99770	.99612

Table 4. $P\{\text{l.u.b.}_{-\infty<y<\infty}|F(y)-F_\lambda^*(y)|<\varepsilon/\sqrt{\bar\lambda}\}$

	λ								
ε	15	20	25	30	35	40	45	50	$\lim_{\lambda\to\infty}$
.4	.011	.004	.001	.006	.006	.004	.003	.002	.0006
.5	.019	.028	.028	.022	.015	.019	.021	.020	.0092
.6	.076	.071	.055	.063	.063	.058	.053	.056	.0414
.7	.141	.128	.132	.127	.123	.125	.124	.119	.1027
.8	.214	.216	.209	.210	.209	.205	.206	.205	.1852
.9	.306	.304	.302	.301	.299	.298	.297	.296	.2776
1.0	.398	.393	.395	.391	.391	.389	.389	.389	.3708
1.1	.480	.482	.477	.480	.476	.477	.475	.475	.4593
1.2	.558	.557	.560	.556	.557	.555	.556	.554	.5404
1.3	.633	.629	.627	.628	.626	.626	.625	.625	.6130
1.4	.689	.690	.693	.689	.688	.689	.687	.688	.6770
1.5	.745	.744	.743	.743	.744	.742	.743	.741	.7328
1.6	.791	.791	.793	.790	.789	.790	.789	.788	.7808
1.7	.829	.829	.829	.829	.830	.829	.828	.829	.8217
1.8	.866	.865	.865	.863	.862	.862	.863	.862	.8563
1.9	.890	.890	.890	.890	.890	.891	.890	.889	.8851
2.0	.913	.914	.915	.914	.913	.912	.912	.913	.9090
2.1	.933	.931	.931	.931	.931	.931	.932	.931	.9285
2.2	.946	.947	.948	.947	.947	.947	.946	.946	.9444
2.3	.959	.959	.958	.958	.958	.958	.958	.959	.9571
2.4	.968	.968	.969	.968	.968	.968	.969	.969	.9672
2.5	.977	.976	.975	.976	.976	.976	.976	.976	.9752
2.6	.981	.981	.982	.982	.982	.982	.982	.982	.9814
2.7	.985	.986	.986	.986	.986	.986	.986	.986	.9861
2.8	.989	.989	.990	.990	.990	.990	.990	.990	.9898
2.9	.992	.992	.992	.992	.992	.992	.992	.992	.9925
3.0	.993	.994	.994	.994	.994	.994	.994	.994	.9946

REFERENCES

1. ALLEN, J. L. and BEEKMAN, J. A. (1966). A statistical test involving a random number of random variables. *Ann. Math. Statist.* **37**, 1305–1311.
2. —— (1967). Distributions of a M. Kac Statistic. *Ann. Math. Statist.* **38**, 1919–1923.
3. BIRNBAUM, Z. W. (1953). On the power of a one-sided test of fit for continuous distribution functions. *Ann. Math. Statist.* **24**, 484–489.
4. BIRNBAUM, Z. W. and TINGEY, F. H. (1951). One-sided confidence contours for probability distribution functions. *Ann. Math. Statist.* **22**, 592–596.
5. CSÖRGÖ, M. and CSÖRGÖ, S. (1970). An invariance principle for the empirical process with random sample size. *Bull. Amer. Math. Soc.* **76**, 706–710.
6. CSÖRGÖ, M. and ALVO, M. (1970). Distribution results and power functions for Kac statistics. *Ann. Inst. Statist. Math.* (Tokyo) **22**, 257–260.
7. DOOB, J. L. (1949). Heuristic approach to the Kolmogorov-Smirnov Theorems. *Ann. Math. Statist.* **20**, 393–403.
8. ERDÖS, P. and KAC, M. (1946). On certain limit theorems of the theory of probability. *Bull. Amer. Math. Soc.* **52**, 292–302.
9. FELLER, W. (1948). On the Kolmogorov-Smirnov limit theorems for empirical distributions. *Ann. Math. Statist.* **19**, 177–189.
10. FERNANDEZ, P. J. (1970). A weak convergence theorem for random sums of independent random variables. *Ann. Math. Statist.* **41**, 710–712.
11. KAC, M. (1949). On deviations between theoretical and empirical distributions. *Proc. Nat. Acad. Sci. U.S.A.* **35**, 252–257.
12. LOÈVE, M. (1960). *Probability Theory*, 2nd Ed. Van Nostrand, Princeton.
13. MASSEY, F. J. (1950). A note on the estimation of a distribution function by confidence limits. *Ann. Math. Statist.* **21**, 116–119.
14. PYKE, R. (1968). The weak convergence of the empirical process with random sample size. *Proc. Cambridge Philos. Soc.* **64**, 155–160.
15. TAKÁCS, L. (1965). Applications of a ballot theorem in physics and in order statistics. *J. Roy. Statist. Soc. Ser B.* **27**, 130–137.

CHAPTER 7

APPLICATIONS IN PHYSICS—FEYNMAN INTEGRALS

7.0 Introduction

Physicists have contributed greatly to the development of Gaussian Markov processes. The investigations of Albert Einstein in 1905 on Brownian motion helped initiate the development of the process which is now frequently called the Wiener process. The Ornstein–Uhlenbeck process was created by physicists to study the velocities of Brownian motion. References for the processes have been given earlier. It is clearly impossible to list even a sizeable fraction of the papers which involve Gaussian Markov processes and physics. However, several other references are the following. S. Chandrasekhar has published a very readable survey article entitled "Stochastic Problems in Physics and Astronomy" [15]. Many of the fourteen papers which appear in the volume [22] pertain to this subject.

This chapter is concerned with only one application of Gaussian Markov processes to physics. Furthermore, only the quantum mechanics subset of physics is considered. However, the subject to be presented, namely Feynman integrals, has inspired a large amount of research on the part of physicists, and mathematicians.

Probabilities in quantum mechanics are calculated using a complex valued function $\psi(\xi, t)$ of space and time which is a solution of Schroedinger's equation

(1) $\quad \dfrac{i}{2} \dfrac{\partial^2 \psi(\xi, t)}{\partial \xi^2} - i V(\xi, t) \psi = \dfrac{\partial \psi}{\partial t}.$

V is the potential involved. Prob. [System subject to potential V is in $[c, d]$ at time $t] = \int_c^d \overline{\psi(\xi, t)} \psi(\xi, t) d\xi$. In 1948 R. P. Feynman used a function space integral to give a solution to (1). [16]. Luckily for mathematicians, he left some questions open. Feynman *postulated* that ψ could be represented as follows:

$$\psi(\xi, t) = \frac{1}{N} \int_{-\infty}^{\infty} \cdots \int_{-\infty}^{\infty} \exp\left[\frac{i}{\hbar} \int_0^t \left\{ \frac{m\dot{x}^2(\tau)}{2} - iV[x(\tau)] \right\} d\tau \right] \prod_{0 \leqslant \tau \leqslant t} dx(\tau)$$

where the integral over $dx(\tau)$ extends over all continuous paths from $(0, 0)$ to (ξ, t), $\dot{x}(\tau) = dx(\tau)/d\tau$, N is a certain normalization constant, $\hbar$ is Planck's constant divided by 2π, and m is mass. Writing

$$\int_0^t \dot{x}^2(\tau)\, d\tau \doteq \sum_{i=1}^n \frac{[x(\tau_i) - x(\tau_{i-1})]^2}{\tau_i - \tau_{i-1}}$$

for

$$\max_{i=1,\ldots,n} [\tau_i - \tau_{i-1}]$$

small, one is led to think of the multivariate normal distribution and the Wiener integral of

$$\exp\left\{\frac{1}{\hbar}\int_0^t V[x(\tau)]\, d\tau\right\}.$$

M. Kac observed this and published a paper in 1949 related to calculating Wiener integrals through partial differential equations (essentially equation (1) without the i) [19]. Gelfand and Yaglom observed this and published a very interesting paper entitled "Integration in Functional Spaces and Its Application to Quantum Physics," in 1959 [17]. Unfortunately, it contained a mistake. They said that one could obtain a complex measure by letting the Wiener kernel have a complex variance.

In 1960, R. H. Cameron pointed out their error [10]. This meant one could not develop Feynman integrals in the usual measure theoretic manner. Cameron then defined a sequential Wiener integral in which the time differences $\tau_i - \tau_{i-1}$ in the kernel were multiplied by a complex parameter. If this parameter was pure imaginary, the result was a Feynman integral. A valuable formula for Feynman integrals was developed. Provided the functional satisfied certain hypotheses, the Feynman integral of $F[x]$ is the Wiener integral of $F[\sqrt{i}\,x]$. In particular, for $V(z)$ and $\sigma(z)$ analytic and not growing too rapidly, the Wiener integral

$$(2) \quad \int_{C[0,t]} \exp\left\{-i\int_0^t V[\sqrt{i}\,x(s) + \xi]\, ds\right\} \sigma[\sqrt{i}\,x(t) + \xi]\, d_w x = \psi(\xi, t)$$

satisfies the Schroedinger equation and the initial condition

$$\lim_{t \to 0+} \psi(\xi, t) = \sigma(\xi).$$

Since then, a great deal of research has been done to enlarge the class of functionals for which Feynman integrals exist. As this chapter evolves, we will give many references. Although these later papers are very important, the sequential integral idea comes closest to Feynman's goal, and has the most intuitive appeal. We will therefore begin with this subject.

7.1. Sequential Feynman integrals

We will start with sequential Feynman integrals using the Brownian motion process.

Let $\tau \equiv [\tau_1, ..., \tau_n]$ be a variable vector of a variable number of dimensions whose components form a subdivision of $[s, t]$ so that $\tau_0 \equiv s < \tau_1 < \tau_2 < ... < \tau_n \equiv t$. Let $\|\tau\| = \max_{j=1, 2, ..., n}(\tau_j - \tau_{j-1})$. Let $\xi \equiv [\xi_1, ..., \xi_{n-1}]$ denote an unrestricted real vector, where n is determined by τ, and let $\xi_0 \equiv x$, $\xi_n \equiv y$, Let $\psi_{\tau, \xi}(\tau_i) = \xi_i$, $i = 0, 1, ..., n$ and $\psi_{\tau, \xi}$ be linear on $[\tau_{i-1}, \tau_i]$.

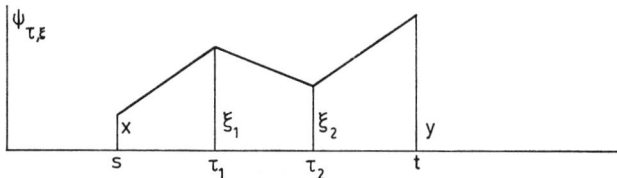

Let $W_{-i}(\tau, \xi)$ be defined as

$$W_{-i}(\tau, \xi) = \frac{\sqrt{2\pi(-i)(t-s)}}{\exp\left\{-\dfrac{[y-x]^2}{2(-i)(t-s)}\right\}} \prod_{i=1}^{n} \frac{\exp\left\{-\dfrac{[\xi_i - \xi_{i-1}]^2}{2(-i)(\tau_i - \tau_{j-1})}\right\}}{\sqrt{2\pi(-i)(\tau_i - \tau_{i-1})}}.$$

Then we define the sequential Feynman integral of the functional $F[X]$ as

$$E^s_{-i}\{F[X] \mid X(s) = x, X(t) = y\} = \lim_{\|\tau\| \to 0} \int_{R_{n-1}} W_{-i}(\tau, \xi) F(\psi_{\lambda, \xi}) d\xi.$$

It is possible to show that you can transform these integrals so that the i's get into the functional $F[X]$. Furthermore, let us now consider the special functional

$$F[X] = \exp\left\{\int_0^t V[X(p) + \xi] dp\right\} \sigma[X(t) + \xi]$$

for an appropriate V and an appropriate σ. Then the sequential Feynman integral of $F[X]$ equals the regular Wiener integral

$$\int_{C[0, t]} \exp\left\{-i \int_0^t V[\sqrt{i}\, X(p) + \xi] dp\right\} \sigma[\sqrt{i}\, X(t) + \xi] d_w X = \psi(\xi, t)$$

and it satisfies the Schroedinger equation and the initial condition

$$\lim_{t \to 0+} \psi(\xi, t) = \sigma(\xi).$$

For general Gaussian Markov processes, our definition of the sequential integral

$$E_\lambda^s\{F[X] \mid X(s) = x\}$$

in Section 4.2 is the one we use here. In this case, we will let

$$G_\lambda(\tau, \xi) = \prod_{i=1}^n p^*(\xi_{i-1}, \tau_{i-1}; \xi_i, \tau_i).$$

Furthermore, subject to the hypotheses of Theorem 2 of that section, we can slightly extend its result to read

$$E_1^s\{F[X] \mid X(s) = x\} = E\{F[X] \mid X(s) = x\}.$$

We will now prove a theorem which relates a sequential integral for complex λ with a regular integral of a modified functional. This presentation is based on [2] and [10].

DEFINITION. Let λ be a fixed non-vanishing complex number such that $-\pi/2 \leqslant \arg \lambda < 0$. The "limiting sequential Gaussian Markov integral with parameter λ" is defined as follows:

(1) $E_\lambda^{\vec{s}}\{F[X] \mid X(s) = x\} = \lim\limits_{\varepsilon \to 0+} E_{\lambda e^{i\varepsilon}}^s\{F[X] \mid X(s) = x\}$

whenever $F[X]$ is a functional such that the right member exists. In particular if $\lambda = (pi)^{-1}$, the "limiting generalized Feynman integral" is defined as follows:

(2) $E_{(pi)^{-1}}^{\vec{f}}\{F[X] \mid X(s) = x\} = \lim\limits_{\varepsilon \to 0+} E_{p^{-1}e^{-i(\pi/2-\varepsilon)}}^s\{F[X] \mid X(s) = x\}$

whenever the right member has meaning.

The word "generalized" is introduced because the original Feynman integrals were for the Wiener process, whereas now we are considering Gaussian Markov processes. For simplicity the above definition restricts the path to a circle arc through λ.

The following theorem gives an existence theorem for complex sequential Gaussian Markov integrals for $-\pi/2 < \arg \lambda < 0$, and for the limiting generalized Feynman integral.

THEOREM 1. Let $p > 0$ and let Λ be the open sector of complex numbers λ such that $-\pi/2 < \arg \lambda < 0$. Let $F(y)$ be a Borel functional defined for all y of the form $\lambda^{-\frac{1}{2}}x(\cdot)$, where $(\lambda, x) \in \Lambda^* x C[s, t]$, and Λ^* denotes the closure of Λ with $\lambda=0$ omitted. Assume that F also satisfies the following four conditions:

1. $F(\lambda^{-\frac{1}{2}}x)$ is analytic in λ throughout Λ for each x in $C[s,t]$.
2. $F(\lambda^{-\frac{1}{2}}x)$ is a continuous function of λ throughout Λ^* for each x in $C[s,t]$.
3. $F(x)$ is a continuous function of x in the uniform topology.
4. For all x in $C[s,t]$ and all γ on $(0, \pi/4)$, $|F(e^{i\gamma}x)| \leq K \exp(M\|x\|)$ where K and M are given integers.

Then the following integrals exist and are equal:

(3) $E_\lambda^s\{F[X] \mid X(s)=0\} = E\{F[\lambda^{-\frac{1}{2}}X] \mid X(s)=0\}$

whenever $\lambda \in S: \{\lambda: \lambda \neq 0, -\pi/2 < \arg \lambda \leq 0\}$.

Moreover, the limiting generalized Feynman integral exists on $C[s,t]$ and

(4) $E_{-i/p}^{\to f}\{F[X] \mid X(s)=0\} = E\{F[(pi)^{\frac{1}{2}}X] \mid X(s)=0\}$.

Finally both members of (3) are continuous throughout S, are analytic inside S, and approach the members of (4) as $\lambda \to -i/p$ from inside S.

REMARK. The proof of this theorem is quite long. However, almost all of the proofs in various papers of the existence of "Feynman integrals" are even longer. Moreover this approach to Feynman's idea is the most intuitive and so it is hoped that the reader will accept the lengthy proof. A key argument is proving that

$$\int_{R_n} G_\lambda(T,z) F(\psi_{T,z}) dz = \int_{R_n} G(T,z) F(\lambda^{-\frac{1}{2}}\psi_{T,z}) dz$$

for complex λ. This is done by analytic continuation, and this is accomplished by showing each of the above quantities is analytic in S, and that equality holds for real λ. The analyticity is demonstrated by Morera's theorem, which requires the use of Fubini's theorem which necessitates showing each of the above quantities is absolutely integrable for $\lambda \in S$. In fact, it requires a little more but the reader will see that as the proof evolves.

Proof. Note from condition 2 that condition 4 holds for $-\pi/2 \leq \gamma \leq 0$. We first show that for each subdivision vector T,

(5) $\int_{R_n} G_\lambda(T,z) F(\psi_{T,z}) dz$

exists for $\lambda \in S$, and is an analytic function of λ throughout S, $\arg \lambda \neq 0$. Note that the integrand of (5) is measurable in z, and that

(6) $|\lambda|^{-n/2}[(2\pi)^n A(s, t_1) \ldots A(t_{n-1}, t_n)]^{\frac{1}{2}} |G_\lambda(T, z) F(\psi_{T,z})|$

$$\leqslant K \exp(M \|\psi_{T,z}\|) \exp\left\{-\operatorname{Re}(\lambda) \sum_{i=1}^n \frac{[z_i - v(t_i) z_{i-1}/v(t_{i-1})]^2}{2A(t_{i-1}, t_i)}\right\}.$$

Let $A^*(a, b) = A(a, b)/\operatorname{Re} \lambda$. We will now show that the above is integrable using Theorem 1 of section 4.1, Theorem 8 of section 4.4, and distribution A of section 4.7.

(7) $\displaystyle\int_{-\infty}^{\infty} (n) \int_{-\infty}^{\infty} K \exp(M\|\psi_{T,z}\|) \exp\left\{-\sum_{i=1}^n \frac{[z_i - v(t_i) z_{i-1}/v(t_{i-1})]^2}{2A^*(t_{i-1}, t_i)}\right\} dz$

$= [(2\pi)^n A^*(s, t_1) \ldots A^*(t_{n-1}, t_n)]^{\frac{1}{2}} E\{K e^{M\|x_T\|} \mid X(s) = 0\}$

$= [(2\pi)^n (\operatorname{Re} \lambda)^{-n} A(s, t_1) \ldots A(t_{n-1}, t_n)]^{\frac{1}{2}}$

$\times \displaystyle\int_{C[0,\frac{u(t)}{v(t)} - \frac{u(s)}{v(s)}]} K \exp\left\{M \sup_{s \leqslant \tau \leqslant t} \left|\frac{v(\cdot)}{\sqrt{\operatorname{Re} \lambda}} X_T\left(\frac{u(\cdot)}{v(\cdot)} - \frac{u(s)}{v(s)}\right)\right|\right\} d_w x$

$\leqslant [(2\pi/\operatorname{Re} \lambda)^n A(s, t_1) \ldots A(t_{n-1}, t_n)]^{\frac{1}{2}} \dfrac{4K}{\sqrt{2\pi\left(\dfrac{u(t)}{v(t)} - \dfrac{u(s)}{v(s)}\right)}}$

$\times \displaystyle\int_0^\infty \exp\left\{\frac{MG}{\sqrt{\operatorname{Re} \lambda}} w - w^2 \Big/ \left[2\left(\frac{u(t)}{v(t)} - \frac{u(s)}{v(s)}\right)\right]\right\} dw \qquad (7)$

where $\qquad v(\tau) \leqslant G,\ s \leqslant \tau \leqslant t$,

and $\qquad \sup_{s \leqslant \tau \leqslant t} |y_T(\tau)| \leqslant \sup_{s \leqslant \tau \leqslant t} |y(\tau)|,$

and $\qquad \sup_{s \leqslant \tau \leqslant t} |y(\tau)| = \max[\sup_{s \leqslant \tau \leqslant t} y(\tau),\ \sup_{s \leqslant \tau \leqslant t} (-y(\tau))]$

and $\qquad \exp\{\sup_{s \leqslant \tau \leqslant t} |y(\tau)|\} \leqslant \exp\{\sup_{s \leqslant \tau \leqslant t} y(\tau)\} + \exp\{\sup_{s \leqslant \tau \leqslant t} (-y(\tau))\}.$

The above quantity is finite.

Now (5) exists for all λ and all subdivision vectors T.

Next let (for each positive integer j)

$$S_j = \{\lambda : \lambda \in S,\ \operatorname{Re} \lambda \geqslant 1/j\}.$$

If we integrate (5) around a contour Γ in S_j, λ will be bounded away from zero on the contour and we can exchange order of integration by the Fubini theorem, since

$$(8) \quad \int_\Gamma \int_{R_n} |G_\lambda(T,z) F(\psi_{T,z})| \, dz \, |d\lambda|$$

is finite by our previous bounds. Because of the analyticity of the integrand of (5) in λ, the repeated integral vanishes, and by Morera's theorem, it follows that (5) is an analytic function of λ inside S_j.

Since $S = \bigcup_{j=1}^\infty S_j$, (5) exists for all λ in S, and all subdivision vectors t, and (5) is an analytic function of λ in S, $\arg \lambda \neq 0$.

For $k=1, 2, ...$, let $S_k = \{\lambda: \lambda \in S, |\lambda|^{-\frac{1}{2}} \geq k\}$. Then for $\lambda \in S_k$, and a subdivision vector t,

$$(9) \quad [(2\pi)^n A(s, t_1) \ldots A(t_{n-1}, t_n)]^{\frac{1}{2}} |G(T, z) F(\lambda^{-\frac{1}{2}} \psi_{T,z})|$$

$$\leq K \exp [Mk \max_{i=1,2,\ldots,n} |z_i|] G(T, z)$$

which is integrable in z.

In a similar fashion then,

$$(10) \quad \int_{R_n} G(T, z) F(\lambda^{-\frac{1}{2}} \psi_{T,z}) \, dz \quad \text{exists for } \lambda \in S_k$$

and is an analytic function of $\lambda \in S_k$, $\arg \lambda \neq 0$.

Since $S = \bigcup_{k=1}^\infty S_k$, (10) exists for all λ in S, and all subdivision vectors t, and (10) is an analytic function of λ in S, $\arg \lambda \neq 0$.

Furthermore (5) and (10) are both continuous throughout S. This follows from (6), (9) and hypothesis 2. Moreover if $\lambda \in S$ and λ is real we may use the fact that

$$G_{p\sigma}(T, z) = p^{n/2} G_\sigma(T, \sqrt{p} z) \text{ for } p > 0$$

and the definition of $\psi_{T,z}$, to find that expression (10) is equal to expression (5) on the real edge of S. By analytic continuation, and the continuity of (5) and (10), we have for all λ in S,

$$(11) \quad \int_{R_n} G_\lambda(T, z) F(\psi_{T,z}) \, dz = \int_{R_n} G(T, z) F(\lambda^{-\frac{1}{2}} \psi_{T,z}) \, dz.$$

Now let λ be real and positive. $F(\lambda^{-\frac{1}{2}} x)$ is Borel measurable and continuous in the uniform topology. Since

$$|F(\lambda^{-\frac{1}{2}} x)| \leq K \exp [M \lambda^{-\frac{1}{2}} \|x\|]$$

and that is finite as before,

(12) $E_1^S\{F[\lambda^{-\frac{1}{2}}x]|x(s)=0\} - E\{F[\lambda^{-\frac{1}{2}}x]|x(s)=0\}$

by a mild extension of Theorem 2 of Section 4.1.

Next let $T^{(1)}$, $T^{(2)}$, ... be a sequence of subdivision vectors such that norm $T^{(i)} \to 0$ as $i \to \infty$ and define

(13) $f_i(\lambda) = \int_{R_{n^{(i)}}} G(T^{(i)}, z) F(\lambda^{-\frac{1}{2}} \psi_{T^{(i)}, z}) dz.$

The functions $f_i(\lambda)$ are defined and continuous for $\lambda \in S$, and are analytic in S, arg $\lambda \neq 0$. Furthermore, from (9) the functions $f_i(\lambda)$ are uniformly bounded for λ in S_k. Moreover from the existence of the left hand side of (12), it follows that for $\lambda > 0$ we have the existence of the limit

(14) $\lim_{i \to \infty} f_i(\lambda) = E\{F[\lambda^{-\frac{1}{2}}x]|x(s)=0\}.$

As in the proof of Theorem 2 [10], it may be proved that (14) holds, throughout S_k, and that both members of (14) are analytic for $\lambda \in S_k$ arg $\lambda \neq 0$, and continuous throughout S_k.

Since

$$S = \bigcup_{k=1}^{\infty} S_k,$$

the above statements apply for S.

Because the limit of $f_i(\lambda)$ is independent of the choice of $T^{(1)}$, $T^{(2)}$, ... it follows from the definition that the sequential Gaussian Markov integral exists and for $\lambda \in S$,

(15) $E_1^S\{F[\lambda^{-\frac{1}{2}}x]|x(s)=0\} = E\{F[\lambda^{-\frac{1}{2}}x]|x(s)=0\}.$

Furthermore from the definition, (11), and (15) the sequential Gaussian Markov integral in (3) exists, is analytic for $\lambda \in S$, arg $\lambda \neq 0$, continuous for $\lambda \in S$, and (3) holds.

Furthermore from hypothesis 4, (2) exists and from hypotheses 1, 2, and 4 we see that the right member of (3) approaches the right member of (2) as $\lambda \to -i/p$ from within S.

7.2. GENERALIZED SCHROEDINGER EQUATIONS

The next theorem shows that for appropriate functionals, the limiting generalized Feynman integral gives a solution to the generalized Schroedinger equation. This theorem has not appeared elsewhere but a different version of it for the Wiener process appears as Theorem 5 in [10].

THEOREM 2. Let $V(u, s)$ be defined over R^*: $0 < s < t$, u any complex number. Let $V(u, s)$ be an entire function of u for each s in $(0, t)$, and such that $V(x, s)$ and $V_x(x, s)$ are continuous and bounded for $-\infty < x < \infty$, $0 \leqslant s \leqslant t$. Also assume that for complex u, $|V(u, s)| \leqslant C|u|$ for some $C > 0$. Assume that $\sigma(x)$ is a complex valued, integrable function of x, with a continuous bounded first derivative for $x \in R$. Also assume $\sigma(u)$ is defined for complex u.

Then it follows that the function $G(\xi, s)$ defined by

(1) $G(\xi, s) = E_{-i}^{\rightarrow f} \left\{ \exp\left\{ -i \int_s^t V[x(p) + \xi v(p)/v(s), p] dp \right\} \right.$

$\left. \times \sigma[x(t) + \xi v(t)/v(s)] \,\middle|\, x(s) = 0 \right\}$

exists and satisfies the partial differential equation

(2) $\dfrac{i}{2}[u'(s) v(s) - u(s) v'(s)] \dfrac{\partial^2 G}{\partial \xi^2} + \xi \dfrac{v'(s)}{v(s)} \dfrac{\partial G}{\partial \xi} - i V G = -\dfrac{\partial G}{\partial s}$

for $0 < s < t$, $-\infty < \xi < \infty$. It also satisfies for each real ξ_0 the boundary condition

(3) $\lim\limits_{\substack{s \to t^- \\ \xi \to \xi_0}} G(\xi, s) = \sigma(\xi_0)$

Proof. The existence of $G(\xi, s)$ follows from Theorem 1. With $p = 1$ in (4) of Theorem 1,

(4) $G(\xi, s) = E\left\{ \exp\left\{ -i \int_s^t V[\sqrt{i}\, x(p) + \xi v(p)/v(s), p] dp \right\} \right.$

$\left. \times \sigma[\sqrt{i}\, x(t) + \xi v(t)/v(s)] \,\middle|\, x(s) = 0 \right\}.$

Now let

(5) $V^*(w, p) = V(\sqrt{i}\, w, p)$ for $0 < s \leqslant p < t$, all complex w, and similarly let

(6) $\sigma^*(w) = \sigma(\sqrt{i}\, w)$ for all complex w.

Let

(7) $G^*(\xi, w, s) = E\left\{ \exp\left\{ -i \int_s^t V^*\left[x(p) + \left(w + \dfrac{\xi}{\sqrt{i}}\right) \dfrac{v(p)}{v(s)}, p \right] dp \right\} \right.$

$\left. \times \sigma^*\left[x(t) + \left(w + \dfrac{\xi}{\sqrt{i}}\right) \dfrac{v(t)}{v(s)} \right] \,\middle|\, x(s) = 0 \right\}.$

If one replaces ξ by $w+\xi/\lambda$ (for a suitable λ neighborhood N of $\sqrt{i}$ which includes an interval on the real axis) in Theorem 5 of section (4.3), the hypotheses allow one to use analytic continuation to obtain for $\lambda = \sqrt{i}$

$$\frac{u'(s)\,v(s) - u(s)\,v'(s)}{2}\frac{\partial^2 G^*}{\partial w^2} + \left(w + \frac{\xi}{\sqrt{i}}\right)\frac{v'(s)}{v(s)}\frac{\partial G^*}{\partial w} + \frac{\partial G^*}{\partial s}$$

$$-iV^*\left(w + \frac{\xi}{\sqrt{i}},\, s\right)G^* = 0$$

for $0 < s < t$, w and ξ real.

Next, observe that the hypotheses allow differentiation within the expectations, and one obtains

$$\frac{\partial G^*}{\partial w} = \sqrt{i}\,\frac{\partial G^*}{\partial \xi},\quad \frac{\partial^2 G^2}{\partial w^2} = i\,\frac{\partial^2 G^*}{\partial \xi^2}.$$

Using these facts and then substituting $w = 0$ gives equation (1) since $G^*(\xi, 0, s) = G(\xi, s)$.

The boundary condition (3) follows from dominated convergence.

Let us now consider some examples of the generalized Schroedinger equation.

Example 1. Wiener process: $u(p) = p$, $v(p) = 1$, $0 \leqslant p \leqslant t$.

$$\frac{i}{2}\frac{\partial^2 G}{\partial \xi^2} - iV(\xi, s)\,G = -\frac{\partial G}{\partial s}$$

Example 2. Doob-Kac process: $u(p) = p$, $v(p) = 1 - p$, $0 \leqslant p \leqslant t < 1$.

$$\frac{i}{2}\frac{\partial^2 G}{\partial \xi^2} - \frac{\xi}{1-s}\frac{\partial G}{\partial \xi} - iV(\xi, s)\,G = -\frac{\partial G}{\partial s}$$

Example 3. Ornstein-Uhlenbeck family of processes:

$$u(p) = \sigma^2 e^{\alpha p},\ v(p) = e^{-\alpha p},\quad \sigma^2 > 0,\ \alpha > 0,\ 0 \leqslant p \leqslant t$$

$$i\sigma^2\alpha\frac{\partial^2 G}{\partial \xi^2} - \alpha\xi\frac{\partial G}{\partial \xi} - iV(\xi, s)\,G = -\frac{\partial G}{\partial s}$$

If we let $f(s) = [u'(s)\,v(s) - u(s)\,v'(s)]/2$, and $g(s) = v'(s)/v(s)$, equation (2) can be written in the form

$$(8)\quad if(s)\frac{\partial^2 G}{\partial \xi^2} + \xi g(s)\frac{\partial G}{\partial \xi} - iVG = -\frac{\partial G}{\partial s}.$$

Letting $g(s) = \frac{1}{2} h(s)$ and

$$V(\xi, s) = \theta(\xi, s) + \tfrac{1}{2} h^2(s)\xi^2 + \frac{i}{2} h(s),$$

equation (8) transforms to

(9) $\quad if(s)\dfrac{\partial^2 G}{\partial \xi^2} + \dfrac{1}{2}\dfrac{\partial}{\partial \xi}[h(s)\,\xi G] - i[\tfrac{1}{2}h^2(s)\,\xi^2 + \theta]G = -\dfrac{\partial G}{\partial s}.$

For $f(s) = \frac{1}{2}$, this is a one-dimensional form of the backward time Schroedinger equation for a "vector" potential $\xi h(s)$ and a potential energy of $\theta(\xi, s)$. See Landau and Lifshitz [21], page 421. The interpretation of a three dimensional forward time version of (9) for the motion of a charged particle in a constant electromagnetic field is given in [17], pages 58–59. An n-dimensional forward time version of (9) for $h(s) = 1$, and ξ replaced by a function of ξ is considered by Babbitt [1]. The corollary on page 805 of [3] explains how to construct a solution to (8) for given $g(s)$, $f(s)$, and $V(\xi, s)$.

7.3. Approximations to Feynman integrals

The approximation of Feynman integrals has been considered by R. H. Cameron [12], S. G. Brush [9], G. Rosen [27], R. L. Zimmerman [28], and others.

The following theorem is part of a theorem which appeared in [5], and which is an extension of Cameron's n-dimensional Riemann integral approximations to Feynman integrals.

We will need the following generic constants (which suppress their dependency on u and v):

$$0 < g \leqslant v(p) \leqslant G, \quad 0 \leqslant p \leqslant T$$

$$d[u(t)/v(t)]/dt \geqslant L > 0, \quad 0 \leqslant t \leqslant T.$$

We will also use another definition of a class of Feynman integrals, studied in [11] and [3]. In this section, and indeed throughout this chapter Feynman integrals could very logically be called Feynman–Cameron integrals, to indicate R. H. Cameron's important contributions to this field.

DEFINITION. Let $F[\varrho x]$ be Lebesgue $\times$ Gaussian integrable in (ϱ, x) on $[0, \varrho_0] x C[s, t]$ for almost all sufficiently small positive ϱ_0 and assume that there exists a function $f(s)$ of bounded variation on $[0, \infty)$ such that for almost all sufficiently large positive λ,

$$\int_0^\infty e^{-s\lambda} df(s) = E\{F[\lambda^{-\frac{1}{2}}x] \mid x(s) = 0\}.$$

Also assume that $f(s)$ is left continuous and $f(0) = 0$. Then $F[x]$ is Feynman integrable on $C[s, t]$ and

$$E^f\{F[x] \mid x(s) = 0\} = \int_0^\infty e^{is} df(s).$$

If

$$\int_0^\infty |e^{s(i-z)}| \, |df(s)| < \infty$$

for all z such that $\mathrm{Re}\, z > 0$ and if the right member of the following exists

$$E^{\to f}\{F[x] \mid x(s) = 0\} = \lim_{\substack{z \to 0 \\ (\mathrm{Re}\, z > 0)}} \int_0^\infty e^{s(i-z)} df(s)$$

we say $F[x]$ is limiting Feynman integrable on $C[s, t]$.

THEOREM 3. Let $\theta(p, u)$ be continuous on the strip $[s, t] x (-\infty, \infty)$, have continuous partial derivatives θ_u and $\theta_{u,u}$ on the strip, and assume

(1) $\displaystyle\int_{-\infty}^\infty |\theta(p, u)| \, du \leqslant B$ for p in $[s, t]$

(2) $\displaystyle\int_{-\infty}^\infty |\theta_{uu}(p, u)| \, du \leqslant B$ on $[s, t]$

(3) $|\theta(p, u)| \leqslant B$ for $(p, u) \in [s, t] x (-\infty, \infty)$.

Let $\sigma(u)$ have two continuous derivatives on $(-\infty, \infty)$ and satisfy

(4) $\displaystyle\int_{-\infty}^\infty |\sigma(u)| \, du \leqslant B$

(5) $\displaystyle\int_{-\infty}^\infty |\sigma''(u)| \, du \leqslant B$

(6) $|\sigma(u)| \leqslant B$ for all u.

Let the functional $F[x]$ be defined by

(7) $F[x] = \exp\left\{\displaystyle\int_s^t \theta[p, x(p)] \, dp\right\} \sigma[x(t)]$

and let

(8) $F_n^*[x] = \Phi_n^*\left\{\int_s^t \theta[p, x(p)] dp\right\} \sigma[x(t)]$

where
$$\Phi_n^*(x) = \sum_{k=0}^{n-1} \frac{x^k}{k!}.$$

Then $F[x]$ and $G_n^*[x]$ are limiting Feynman integrable and the following error estimate applies:

(9) $E^{\to f}\{F[x] \mid x(s) = 0\} = E^{\to f}\{F_n^*[x] \mid x(s) = 0\} + O\left(\left[\frac{5CU(t)}{nv(t)}\right]^n\right)$

where

(10) $C = \max[B/(gL), BG/L, B/L, B/g, BG, B]$.

In particular, if $n \geq 2\sqrt{3}\, CU(t)/v(t)$ (and $n > 1$),

(11) $|E^{\to f}\{F[x] - F_n^*[x]\} \mid x(s) = 0\}|$
$$< \frac{2C^{n+1}[U(t)]^n 3^{(n-1)/2}}{n!\,[v(t)]^n}\left[9\left(\frac{U(t)}{v(t)}\right)^{\frac{1}{2}} + 4n\right].$$

$E^{\to f}\{F_n^*[x] \mid x(s) = 0\}$ can be expressed in terms of iterated Riemann integrals by ($p^*(x, s; \alpha, \tau)$ uses $\lambda = -i$)

(12) $E^{\to f}\{F_n^*[x] \mid x(s) = 0\}$
$$= \int_{-\infty}^{\infty} \sigma(x)\, p^*(0, s; x, t)\, dx + \sum_{k=1}^{n-1} \int_s^t dp_k \int_s^{p_k} dp_{k-1} \ldots \int_s^{p_3} dp_2 \int_s^{p_2} dp_1$$
$$\times \int_{-\infty}^{\infty} {}^{(k+1)} \cdots \int_{-\infty}^{\infty} \prod_{j=1}^{k}[\theta(p_j, x_j)\, p^*(x_{j-1}, \tau_{j-1}; x_j, \tau_j)]$$
$$\times \sigma(x_{k+1})\, p^*(x_k, \tau_k; x_{k+1}, t)\; dx_1 \ldots dx_{k+1}$$

where $p_0 = s$, $x_0 = 0$.

There is one simple situation in which the limiting Feynman integral can be evaluated exactly.

LEMMA. Let $s < p_1 < p_2 < \ldots < p_n \leq t$ and let $\sigma_1(u), \ldots, \sigma_n(u)$ be real or complex functions that have two continuous derivatives and satisfy conditions (4), (5) and (6).

Let
$$G[x] = \prod_{j=1}^{n} \sigma_j[x(t_j)].$$

Then $G[x]$ is limiting Feynman integrable and

(13) $E^{\to f}\{G[x]\,|\,x(s)=0\}$

$$= \int_{-\infty}^{\infty} \overset{(n)}{\cdots} \int_{-\infty}^{\infty} \prod_{j=1}^{n} \{\sigma_j(z_j)\, p^*(z_{j-1},\,t_{j-1};\,z_j,\,t_j)\}\ dz_1\ldots dz_n.$$

7.4. Analytic Feynman integrals and examples

Unfortunately, the previous approximation is not too practical. Moreover, if one studies quantum mechanics textbooks, it is obvious that the hypotheses that we have imposed on the potential $V(y, p)$ are far too strict. Thus, our boundedness eliminates the harmonic oscillator $V(y, p) = y^2$. It therefore would seem that one should calculate $E\{F[\lambda^{-\frac{1}{2}}x]\,|\,x(s)=0\} = J(\lambda)$ for various potentials, see if the result is true for complex λ, and then see if by brute force one can differentiate $J(-i)$ with respect to ξ and t and see if it satisfies the generalized Schroedinger equation. So that this is not all very heuristic, let us make the following

DEFINITION. Let the complex number λ_0 satisfy $\operatorname{Re} \lambda_0 \geqslant 0$, $\lambda_0 \neq 0$ so that $\lambda_0 = |\lambda_0| \exp(i\theta)$ for some θ in $[-\pi/2, \pi/2]$. Let $F[x]$ be a functional defined on $C[s, t]$ and such that

$$J(\lambda) = E\{F[\lambda^{-\frac{1}{2}}x]\,|\,x(s)=0\}$$

exists for all real λ in the interval $|\lambda_0| < \lambda < |\lambda_0| + \delta$, for some $\delta > 0$. Then if $J(\lambda)$ can be extended so that it is defined and continuous on the closed region

$$S = \{\lambda:\, \lambda = re^{i\gamma},\ |\lambda_0| \leqslant r \leqslant |\lambda_0| + (1-\gamma\theta^{-1})\delta,\ \gamma \in [0,\theta]\ \text{or}\ \gamma \in [\theta, 0]\}$$

and analytic in its interior, we define

(1) $E_{\lambda_0}^{\text{ang}}\{F[x]\,|\,x(s)=0\} = J(\lambda_0)$,

and we call the left member of (1) "the analytic Gaussian integral of $F[x]$ with parameter λ_0".

In particular, if $\lambda_0 = -i/p$ where $p > 0$, the integral (1) will be called "the analytic generalized Feynman integral with parameter p", and we write

(2) $E_p^{\text{anf}}\{F[x]\,|\,x(s)=0\} = E_{-i/p}^{\text{ang}}\{F[x]\,|\,x(s)=0\}$.

We will now consider three examples which appear in the discussion of motion in one dimension in Landau and Lifshitz [21], pages 60–80. For simplicity, we assume that $\hbar=1$ and mass is equal to one.

Example 1. Zero potential.

$$V(y,p) \equiv 0, \quad -\infty<y<\infty,\ s\leqslant p\leqslant t$$

(3) $\sigma(y) = \exp\{-y^2/(4b^2)\}[2\pi b^2]^{-\frac{1}{4}}, \quad -\infty<y<\infty,\ b>0.$

Then $\sigma^2(y)$ is the normal density with mean 0 and variance b^2.

This approximates a Dirac delta function as $b\to 0$, because as $b\to 0$ the probability becomes concentrated at the point 0.

From (13) of section 7.3,

(4) $E^{\to f}\{F[x]\,|\,x(s)=0\}$

$$= \frac{\exp(-\pi i/4)}{[2\pi A(s,t)]^{\frac{1}{2}}} \int_{-\infty}^{\infty} \sigma\!\left[x+\xi\frac{v(t)}{v(s)}\right] \exp\!\left[\frac{ix^2}{2A(s,t)}\right] dx$$

$$= \frac{(2b^2)^{\frac{1}{4}}}{\{\sqrt{\pi}[2b^2+iA(s,t)]\}^{\frac{1}{2}}} \exp\!\left\{\frac{-\xi^2 v^2(t)}{2v^2(s)[2b^2+iA(s,t)]}\right\}.$$

Now consider the Wiener process for which $v(t)\equiv 1$ and $A(0,t)=t$, and let $\psi(\xi,t)$ equal the above integral. One can then verify that it satisfies the Schroedinger equation

$$\frac{\partial \psi}{\partial t} = \frac{i}{2}\frac{\partial^2 \psi}{\partial \xi^2} + V\psi$$

with initial condition

$$\lim_{t\to 0+}|\psi(\xi,t)|^2 = \sigma^2(\xi), \quad -\infty<\xi<\infty.$$

For a system $x(s)$, $0\leqslant s\leqslant t$ subject to $V(y)\equiv 0$ and above initial condition, the quantum mechanical method of calculating probabilities yields the equation

$$\text{Prob}\{c\leqslant x(t)\leqslant d\} = \int_c^d \psi(\xi,t)\,\overline{\psi(\xi,t)}\,d\xi$$

$$= \int_c^d \frac{b}{[2\pi(b^4+t^2/4)]^{\frac{1}{2}}} \exp\!\left\{\frac{-\xi^2 b^2}{2(b^4+t^2/4)}\right\} d\xi.$$

In this case $\psi(\xi, t)\overline{\psi(\xi, t)}$ is, for each fixed t, a normal density with mean 0 and variance $(b^4 + t^2/4)/b^2$. Hence, for each fixed t,

$$\int_{-\infty}^{\infty} \psi(\xi, t) \cdot \overline{\psi(\xi, t)}\, d\xi = 1.$$

Let us consider the family of densities as t grows. The variance is growing as t grows, forcing the density to spread out. For example, the variance is b^2 when $t=0$, but at $t=2b^2$, the variance is $2b^2$.

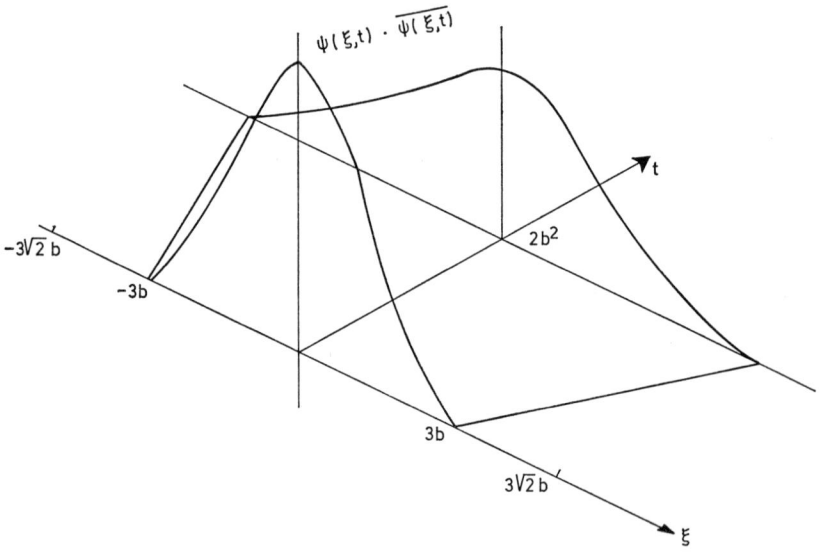

The fact that the means all equal zero reflects the central tendency of the particles.

Example 2. Motion in a homogeneous field. (See [21], page 73.)

Let $\sigma(y)$ be given by (3) and $V(y) = (A + By)$, A and B real, $-\infty < y < \infty$.

One can compute $J(\lambda)$, using the Cameron-Martin translation theorem for Wiener integrals. See page 262 of [5]. $J(\lambda)$ is analytic in a suitable neighborhood, so $J(-i)$ gives the analytic Feynman integral.

For the Wiener process, with $s=0$,

$$(5) \quad J(-i) = \psi(\xi, t) = \frac{(2b^2)^{\frac{1}{4}}}{[\sqrt{\pi}(2b^2 + it)]^{\frac{1}{2}}}$$

$$\exp\left\{-it(A + B\xi) - \frac{iB^2 t^3}{6} - \frac{(\xi + Bt^2/2)^2}{[2(2b^2 + it)]}\right\}.$$

One can easily verify that

$$\frac{i}{2}\frac{\partial^2 \psi}{\partial \xi^2} - i(A+B\xi)\psi = \frac{\partial \psi}{\partial t}$$

and that
$$\lim_{t \to 0+} \psi(\xi, t) = \sigma(\xi), \quad -\infty < \xi < \infty.$$

In this example,

$$\psi(\xi, t) \cdot \overline{\psi(\xi, t)} = \frac{b}{[2\pi(b^4 + t^2/4)]^{\frac{1}{2}}} \exp\left\{\frac{-b^2(\xi + Bt^2/2)^2}{2(b^4 + t^2/4)}\right\}$$

which is a normal density with mean $-Bt^2/2$ and variance $b^2/(b^4 + t^2/4)$. It is curious that A is not involved. The density $\psi \cdot \bar{\psi}$ achieves its maximum at $\xi = -t^2B/2$ which is the point where a classical particle acted on by this potential would be.

For ease of graphing, let us assume that B is negative. Then as t grows, both the mean and variance are increasing.

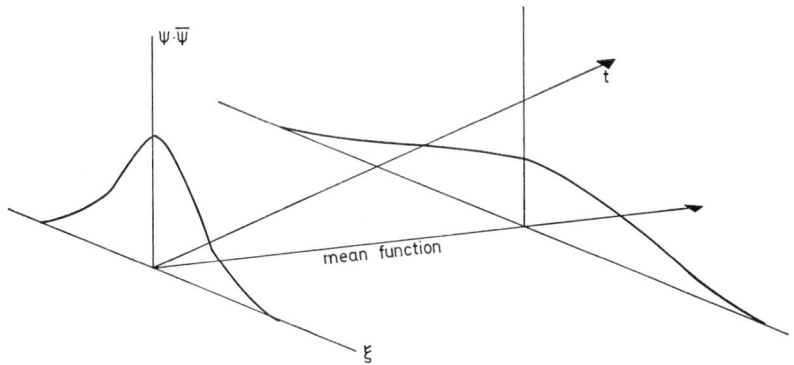

Example 3. Harmonic oscillator. (See [21], page 67.)

Let $\sigma(y)$ be given by (3) and $V(y) = Cy^2$. On pages 263–264 of [5], $J(\lambda)$ is calculated for the Wiener process. The methods are similar to those used in Example 7 of Section 4.1 and also draw on Chapter 4's reference [1] by Donsker and Lions. With $J(-i) = \psi(\xi, t)$,

$$\psi(\xi, t) = \left(\frac{Cb^2}{\pi}\right)^{\frac{1}{4}} \exp\left\{\frac{-\xi^2[-i\sqrt{2C}\cot\sqrt{2C}\,t + 4Cb^2]}{2 - 4ib^2\sqrt{2C}\cot\sqrt{2C}\,t}\right\}$$

$$\times \left\{\frac{-2i}{-2ib^2\sqrt{2C}\cos\sqrt{2C}\,t + \sin\sqrt{2C}\,t}\right\}^{\frac{1}{4}}$$

One can verify that

$$\frac{i}{2}\frac{\partial^2 \psi}{\partial \xi^2} - iC\xi^2\psi = \frac{\partial \psi}{\partial t}, \quad \text{and} \quad \lim_{t\to 0+} \psi(\xi, t) = \sigma(\xi), \quad -\infty < \xi < \infty.$$

In this case,

$$\psi(\xi, t) \cdot \overline{\psi(\xi, t)} = \left(\frac{4Cb^2}{\pi}\right)^{\frac{1}{2}} [8Cb^4 \cos^2 \sqrt{2C}\,t + \sin^2 \sqrt{2C}\,t]^{-\frac{1}{2}}$$

$$\times \exp[-4Cb^2\xi^2(8Cb^4 \cos^2 \sqrt{2C}\,t + \sin^2 \sqrt{2C}\,t)^{-1}]$$

which is a normal density with a time dependent variance which oscillates.

The frequency of oscillation of the Gaussian distribution $\psi\bar{\psi}$ is equal to the frequency of a classical harmonic oscillator with the force constant C.

Since the mean function $\equiv 0$, $\psi\bar{\psi}$ does not move off on a ray as in Example 2.

However, the variance (for $C=1$)

$$= \frac{[8b^4 \cos^2 \sqrt{2}\,t + \sin^2 \sqrt{2}\,t]}{8b^2}$$

oscillates as t grows. This is partially exhibited in the following brief table:

t	Variance
0	b^2
$\frac{\pi}{2\sqrt{2}}$	$\frac{1}{8b^2}$
$\sqrt{2}\,\pi$	b^2

7.5. Forward time equations

The reader may have observed that the generalized Schroedinger equations presented in section 7.2 involved a backwards time derivative. In [6] a relevant theorem appears. Let $p^*(x, s; y, t)$ denote the transition density function with $A(s, t)$ replaced by $A(s, t)/\lambda$ for $\lambda > 0$. Let $E^*\{F[X] \mid X(s) = x, X(t) = y\}$ denote the expected value of a functional $F(X)$ for the stochastic process with that transition density function.

THEOREM 1. Assume $V(y, \tau)$ and $V_y(y, \tau)$ are bounded and continuous complex valued functions for $0 \leq \tau \leq T < \infty$, $y \in \{(-\infty, \infty) - E\}$ where E is a finite set of points. Assume $\lambda > 0$. Then

$$r^*(x, s; y, t) = E^* \left\{ \exp\left[-i \int_s^t V[X(\tau), \tau] d\tau \right] \Big| X(s) = x, X(t) = y \right\} p^*(x, s; y, t)$$

uniquely satisfies the pair of equations

(1) $$\frac{A(t)}{\lambda} \frac{\partial^2 r^*}{\partial y^2} - B(t) \frac{\partial}{\partial y}[yr^*] - iV(y, t) r^* = \frac{\partial r^*}{\partial t}$$

and

(2) $$\frac{A(s)}{\lambda} \frac{\partial^2 r^*}{\partial x^2} + xB(s) \frac{\partial r^*}{\partial x} - iV(x, s) r^* = -\frac{\partial r^*}{\partial s},$$

where

(3) $A(t) = [v(t) u'(t) - u(t) v'(t)]/2$, $0 \leq t \leq T$,

(4) $B(t) = v'(t)/v(t)$, $0 \leq t \leq T$.

It also satisfies the boundary conditions

(5) $$\lim_{t \to s+} \int_{-\infty}^{\infty} g(x) r^*(x, s; y, t) dx = g(y)$$

for every bounded continuous g, and

(6) $$\lim_{s \to t-} \int_{-\infty}^{\infty} g(y) r^*(x, s; y, t) dy = g(x)$$

for every bounded continuous g, $\lim_{|y| \to \infty} r^* = 0$, $\lim_{|x| \to \infty} r^* = 0$ and the continuity properties $\partial r^*/\partial y$, $\partial^2 r^*/\partial y^2$, $\partial r^*/\partial t$ continuous for $0 \leq s < t < T$, $y \in \{(-\infty, \infty) - E\}$ and $\partial r^*/\partial x$, $\partial^2 r^*/\partial x^2$, $\partial r^*/\partial s$ continuous for $0 < s < t \leq T$, $x \in \{(-\infty, \infty) - E\}$ and $\partial r^*/\partial y$ continuous for $y \in E$, $\partial r^*/\partial x$ continuous for $x \in E$.

This theorem has the weakness that the parameter λ must be positive. However, it suggests that if one were to calculate the expected value, and if the expression persisted for complex λ, that by brute force one might be able to show that it satisfied equations (1) and (2). One is now faced with the problem that it is very difficult to calculate function space integrals tied down at the right time extremity. For this situation, Theorem 4 of section 4.2 is helpful.

We will now consider two examples which are discussed in [7].

Example 1. Zero Potential: $V(y, \tau) \equiv 0$.

$$E_\lambda^s\{1 \mid X(s) = x, X(t) = y\} p^*(x, s; y, t)$$

$$= \frac{1}{2\pi} \int_{-\infty}^{\infty} \exp\left\{-i\mu\left[y - x\frac{v(t)}{v(s)}\right]\right\} E\{\exp(i\mu X(t)/\sqrt{\lambda}) \mid X(s) = 0\} d\mu$$

$$= \frac{1}{2\pi} \int_{-\infty}^{\infty} \exp\left\{-i\mu\left[y - x\frac{v(t)}{v(s)}\right]\right\} \int_{-\infty}^{\infty} \exp[i\mu y/\sqrt{\lambda}] p(0, s; y, t) \, dy \, d\mu$$

$$= \frac{1}{2\pi} \int_{-\infty}^{\infty} \exp\left\{-i\mu\left[y - x\frac{v(t)}{v(s)}\right]\right\} \exp\left\{-\tfrac{1}{2} A(s, t) \mu^2/\lambda\right\} d\mu$$

by the Lemma on p. 144 of [6]

$$= p^*(x, s; y, t).$$

For Re $\lambda = 0$, one also obtains $p^*(x, s; y, t)$ by using Fresnel integrals. One can now substitute $\lambda = -i$, and verify that the result satisfies (1) and (2). It does not seem possible to verify the boundary conditions (5) and (6) for all bounded continuous g's, but conditions (5) and (6) were verified for $g(x) \equiv 1$, and $g(x) = \exp(-x^2/2)$, $-\infty < x < \infty$.

Example 2. Forced Harmonic Oscillator:

$$V(y, \tau) = y^2 - f(\tau) y$$

where $\qquad f(0) = 0, \quad f(T) = 0$

and $\qquad f(\tau) \in L_2[0, T]$.

The value of the integral in the Wiener case and for $\lambda = -i$ was recorded in section 4.2. The result satisfies equations (1) and (2) for the Wiener process and conditions (5) and (6), thus justifying calling $r^*(x, s; y, t)$ a Green's function for the Schroedinger equations.

REMARK. It would be interesting to let the forcing function be a combination of $\sin kt$ and $\cos kt$ with the same oscillation as the variance in Example 3 of section 7.4.

7.6. CAPSULE VIEW OF SOME REFERENCES

In this section, brief mention will be made of two papers, and a more lengthy discussion of three other papers will be made.

The connection between Gaussian Markov stochastic processes and generalized Schroedinger equations has been discussed in [26], with heavy emphasis on the physics involved.

In [13] and [14], R. H. Cameron and D. A. Storvick studied an operator valued function space integral which satisfied the following integral equation:

$$\psi(\xi, t) = \underset{\substack{A \to \infty \\ \xi}}{\text{l.i.m.}} \left(\frac{1}{2\pi i t}\right)^{\frac{1}{2}} \int_{-A}^{A} \sigma(u) \exp\left\{\frac{i(\xi - u)^2}{2t}\right\} du$$

$$+ \underset{\substack{A \to \infty \\ \xi}}{\text{l.i.m.}} \left(\frac{1}{2\pi i}\right)^{\frac{1}{2}} \int_{0}^{t} \int_{-A}^{A} \frac{1}{\sqrt{t-s}} V(u, s) \psi(u, s) \exp\left\{\frac{i(\xi - u)^2}{2(t-s)}\right\} du \, ds.$$

If you ignore the limits, you can heuristically differentiate under the integral signs and verify that the $\psi(\xi, t)$ of the integral equation does satisfy the Schroedinger equation:

$$\frac{\partial \psi}{\partial t} = \frac{i}{2} \frac{\partial^2 \psi}{\partial \xi^2} + V\psi.$$

Since references [13] and [14] impose fewer restrictions on V, one can say that Cameron and Storvick enlarged the class of potentials for which the Schroedinger equation has a solution in terms of a Feynman integral.

The analogous integral equations for Gaussian Markov processes were derived by Ralph Kallman and the author in [8]. Reference [7] also relates to this subject.

Feynman's idea of a path integral has been used outside the realm of physics. For example in the Bucy–Joseph book mentioned in chapter 1, on page 43 this idea is utilized. There the conditional distribution of the signal given the observations is represented as the expectation of a functional of the observations. This distribution solves the filtering problem in the non-linear case. To quote Bucy and Joseph: "Further dynamical equations for the optimal filter can be easily derived from the representation theorem. The representation theorem is the fundamental part of the theory of filtering from which all the theory can be derived. It is analogous to the "path integral" approach of Feynman for the solution of the Schroedinger equation in quantum theory and mathematically it is the explicit formula for the Radon–Nikodym derivative of one function space measure with respect to another."

7.7. FINITE DIFFERENCE APPROACH TO GENERALIZED SCHROEDINGER EQUATIONS

One of the frustrating things about Feynman integrals is that they are even harder to calculate than Wiener integrals. Some approximations have been developed; for example, the n-dimensional Riemann integral approximations of R. H. Cameron (see [12]). However, the results are still not too practical for computation.

Another way to approximate Feynman integrals is to approximate the generalized Schroedinger equation by a difference scheme. This method assumes that the potential $V(x, t)$ and initial condition $\sigma(x)$ (if involved) satisfy the conditions of one of the existence theorems. The pleasant fact is that even when the $V(x, t)$ and $\sigma(x)$ don't satisfy the conditions, probably the Feynman integral exists, especially if the difference scheme method performs satisfactorily. This is exactly what takes place in two of the examples of section 7.4.

Assume that as a function of y and t

(1) $\quad E_\lambda^s \left\{ \exp\left[-i \int_s^t V(X(\tau), \tau) d\tau \right] \middle| X(s) = x, X(t) = y \right\} p^*(x, s; y, t)$

$$= G(y, t) + iH(y, t)$$

and assume that G and H have derivatives as indicated below. Substituting this in the forward generalized Schroedinger equation and equating real and imaginary parts leads to the two partial differential equations

(2) $\quad A(t) \dfrac{\partial^2 G}{\partial y^2} - B(t) y \dfrac{\partial H}{\partial y} - B(t) H - VG = \dfrac{\partial H}{\partial t}$

(3) $\quad -A(t) \dfrac{\partial^2 H}{\partial y^2} - B(t) y \dfrac{\partial G}{\partial y} - B(t) G + VH = \dfrac{\partial G}{\partial t}$

where $A(t)$ and $B(t)$ are defined in section 7.5. As an approximation to the Dirac delta function initial condition, let

(4) $\quad G(y, 0+) + iH(y, 0+) = [e^{-y^2/(2\sigma^2)}/(2\pi\sigma^2)]^{\frac{1}{2}}, \quad \sigma^2 > 0.$

As $\sigma^2 \to 0+$, the desired initial condition is produced.

The unboundedness of the initial condition precludes proofs of convergence and stability of difference schemes, but electronic computer programs were run and proved fairly successful for moderately long t's. Because of their success, an indication will be made of some of the difference schemes used. They were motivated by two M. S. theses, [18], [23], which considered the Wiener process for initial condition (4) with $\sigma^2=1$. The first thesis used an explicit scheme and obtained convergence and stability for $(\Delta t) \leq (\Delta y)^3$. The second thesis used an implicit scheme and obtained convergence and stability for $(\Delta t) = (\Delta y)^2/(2+\varepsilon)$, $\varepsilon > 0$. These proofs could be generalized to the Gaussian Markov case for each fixed value of σ^2. Those theses considered $V(y, t) \equiv 0$, and

$$V(y, t) = \begin{cases} .1\, y^2, & |y| \leq 7 \\ 0, & |y| > 7 \end{cases}$$

and

$$G(y, 0+) = \begin{cases} e^{-y^2/4}/(2\pi)^{1/4}, & |y| \leq 7 \\ 0, & |y| > 7 \end{cases}.$$

Electronic computer programs were run and sample probabilities of the form Prob {System subject to potential V and initial condition $G(y, 0+)$ is in $[c, d]$ at time t} $= \int_c^d \psi(\xi, t) \cdot \overline{\psi(\xi, t)}\, d\xi$ were calculated. The exact answers to these probabilities were available through section 7.4. The errors for the sample probabilities were less than .005 for $t \leq 10$. Because of the choice of units, namely that $\hbar = 1$ and mass is 1, $t = 10$ would be a reasonably long period.

The grid values were $y = jh$, $t = nk$, for $h = .2$, $k = .005$. Values of σ^2 as small as .001 were used. For the Wiener case $A(nk) = \frac{1}{2}$ and $B(nk) = 0$. Converting the derivatives into difference quotients and designating time by a superscript n and the spatial position by a subscript j produced the equations

(5) $H_j^{n+1} = H_j^n + \dfrac{k}{2h^2}(G_{j+1}^{n+1} - 2G_j^{n+1} + G_{j-1}^{n+1}) - kVG_j^n$

(6) $G_j^{n+1} = G_j^n - \dfrac{k}{2h^2}(H_{j+1}^{n+1} - 2H_j^{n+1} + H_{j-1}^{n+1}) + kVH_j^n.$

For the Ornstein-Uhlenbeck case, $A(nk) = 1$, and $B(nk) = -1$. Here the equations used were

(7) $H_j^{n+1} = H_j^n + \dfrac{k}{h^2}(G_{j+1}^{n+1} - 2G_j^{n+1} + G_{j-1}^{n+1}) + kj(H_{j+1}^{n+1} - H_{j-1}^{n+1}) + kH_j^n - kVG_j^n$

(8) $G_j^{n+1} = G_j^n - \dfrac{k}{h^2}(H_{j+1}^{n+1} - 2H_j^{n+1} + H_{j-1}^{n+1}) + kj(G_{j+1}^{n+1} - G_{j-1}^{n+1}) + kG_j^n + kVH_j^n.$

The five potentials for which computer programs were run were $V(y, t) = 0$, $V(y, t) = .1\, y^2$, $V(y, t) = .1\, y^2 - .1\, y \sin(\pi t)$, $V(y, t) = .1\, y^2/(1+t)^2$, $V(y, t) =$

$-1/(f(|y|, t)+|y|)^2$. The last two potentials were motivated by references [24] and [25] respectively. Analytic expressions for ψ in the first three and fifth cases may be found in [5], [6], and [24]. An analytic expression for the fourth ψ could be calculated as in [6], using a modified version of the representation of the process as in [20], page 200. Potentials two, three, four, and five all are unbounded, thus not fulfilling one of the hyptheses of the existence theorem for solutions to the generalized Schroedinger equations. However, in the first three cases the rapid growth occurs where the initial condition is very small, and furthermore, the ψ values were cut off at $|y|=7$. The desired potential five, $V(y, t) = -1/y^2$, ran into trouble because its maximum values occurred at the maximum values of the initial condition. Hence the convergence factor $f(|y|, t)$ was added. For the Wiener process, $f(|y|, t) = 1.0 + .120\,(t-100)/100 - .143\,|y| - .017\,(t-100)|y|/100$ for $t=100$ (100) 1 200, $|y| \leqslant 7$. For the Ornstein–Uhlenbeck process, $f(|y|, t) = .75 + .120\,(t-100)/100 - .107\,|y| - .017\,(t-100)\,|y|/100$ for $t=100$ (100) 1 200, $|y| \leqslant 7$.

REFERENCES

1. BABBITT, D. G. (1965). The Wiener integral and the Schroedinger operator. *Trans. Amer. Math. Soc.* **116**, 66–78. (Correction, *Trans. Amer. Math. Soc.* **121**, 549–552 (1966).)
2. BEEKMAN, J. A. (1963). Solutions to generalized Schroedinger equations via Feynman integrals connected with Gaussian Markov stochastic processes. Ph. D. Thesis. Univ. of Minnesota.
3. — (1965). Gaussian processes and generalized Schroedinger equations. *J. Math. Mech.* **14**, 789–806.
4. — (1967). Gaussian Markov processes and a boundary value problem. *Trans. Amer. Math. Soc.* **126**, 29–42.
5. — (1967). Feynman-Cameron integrals. *J. Math. Physics* **46**, 253–266.
6. — (1969). Green's functions for generalized Schroedinger equations. *Nagoya Math. J.* **35**, 133–150. (Correction: *Nagoya Math. J.* **39** 199 (1970).
7. — (1971). Sequential Gaussian Markov integrals. *Nagoya Math. J.* **42**, 9–21.
8. BEEKMAN, J. A. and KALLMAN, R. A. (1971). Gaussian Markov expectations and related integral equations. *Pacific Math. J.* **37**, 303–317.
9. BRUSH, S. G. (1961). Functional integrals and statistical physics. *Rev. Mod. Physics* **33**, 79–92.
10. CAMERON, R. H. (1960). A family of integrals serving to connect the Wiener and Feynman integrals. *J. Math. Physics* **39**, 126–140.
11. — (1962–63). The Ilstow and Feynman integrals. *J. d'Analyse Mathematique* **10**, 287–361.
12. — (1968). Approximations to certain Feynman integrals. *J. d'Analyse Mathematique* **21**, 337–371.
13. CAMERON, R. H. and STORVICK, D. A. (1968). An operator valued function space integral and a related integral equation. *J. Math. Mechanics* **18**, 517–552.
14. — (1970). An integral equation related to the Schroedinger equation with an application to integration in function space. Bochner Memorial Volume, pp. 175–193. Princeton University Press, Princeton, N.J.
15. CHRANDRASEKHAR, S. (1943). Stochastic problems in physics and astronomy. *Rev. Mod. Physics* **15**, 3–91. Reprinted in *Selected Papers on Noise and Stochastic Processes* (1954), edited by Nelson Wax. Dover, New York.
16. FEYNMAN, R. P. (1948). Space-time approach to non-relativistic quantum mechanics. *Rev. Mod. Physics* **20**, 367–387.
17. GELFAND, I. M. and YAGLOM, A. M. (1960). Integration in functional space and its applications in quantum physics. *J. Math. Physics* **1**, 48–69.

18. HUMMEL, K. G. (1966). An error analysis of Morgan's method, and an alternate method for approximating the Schroedinger wave equation. M. S. Thesis. Ball State University.
19. KAC, M. (1949). On distributions of certain Wiener functionals. *Trans. Amer. Math. Soc.* **65**, 1–13.
20. — (1951). On some connections between probability theory and differential and integral equations. *Proc. Second Berkeley Symposium*, pp. 189–215. University of California Press, Berkeley, California.
21. LANDAU, L. and LIFSHITZ, E. (1958). *Quantum Mechanics, Non-relativistic Theory*. Addison-Wesley, Reading, Mass.
22. MARTIN, W. T. and SEGAL, I. Editors (1964). *Analysis in Function Space*. M.I.T. Press, Cambridge, Mass.
23. MORGAN, T. J. (1966). A numerical solution of the Schroedinger wave equation. M. S. Thesis. Ball State University.
24. NELSON, E. (1964). Feynman integrals and the Schroedinger equation. *J. Math. Physics*. **5**, 332–343.
25. PEÑA-AUERBACH, L. DE LA, BRAUN, E. and GARCIA-COLIN, L. S. (1968). Quantum-mechanical description of a Brownian particle. *J. Math. Physics* **9**, 668–674.
26. PEÑA-AUERBACH, L. DE LA and GARCIA-COLIN, L. S. (1968). Simple generalization of Schroedinger's equations. *J. Math. Physics* **9**, 922–927.
27. ROSEN, G. (1963). Approximate evaluation of Feynman functional integrals. *J. Math. Physics* **4**, 1327–1333.
28. ZIMMERMAN, R. L. (1965). Evaluation of Feynman's functional integrals. *J. Math. Physics* **6**, 1117–1124.

ANSWERS TO SELECTED EXERCISES

CHAPTER 1

1. For part c, use L'Hospital's rule.

$$\lim_{t\to 0+} \frac{1}{\sqrt{2\pi 2Dt}} \exp\left\{-\frac{(x-x_0)^2}{4Dt}\right\}$$

$$= \lim_{t\to 0+} \frac{t^{-\frac{1}{2}}}{\sqrt{4\pi D}} \frac{1}{\exp\left\{\frac{(x-x_0)^2}{4Dt}\right\}}$$

$$= \lim_{t\to 0+} \frac{(-\frac{1}{2})t^{-\frac{3}{2}}}{\sqrt{4\pi D}} \frac{1}{\exp\left\{\frac{(x-x_0)^2}{4Dt}\right\} \frac{(x-x_0)^2}{4D}(-1)t^{-2}}$$

$$= \lim_{t\to 0+} \frac{(\frac{1}{2})t^{\frac{1}{2}}}{\sqrt{4\pi D}} \frac{1}{\exp\left\{\frac{(x-x_0)^2}{4Dt}\right\} \frac{(x-x_0)^2}{4D}}$$

$$= 0.$$

2. $P\{-1 \leqslant x(t) \leqslant 1, \ 0 \leqslant t \leqslant 1\}$

$$\doteq \frac{4}{\pi}\left\{e^{-\pi^2/16} - \frac{1}{3}e^{-9\pi^2/16} + \frac{1}{5}e^{-25\pi^2/16}\right\}$$

$$\doteq 1.2732\{.5395 - .0013 + .0000\}$$

$$= .6852.$$

With only two terms, the error was $\leqslant .001$.

3. $P[\sup_{0\leqslant t\leqslant 1} x(t) \geqslant 1] = \frac{\sqrt{2}}{\sqrt{\pi}}\int_1^\infty e^{-y^2/2}dy$

$$= 2\left[1 - \frac{1}{\sqrt{2\pi}}\int_{-\infty}^1 e^{-y^2/2}dy\right]$$

$$= 2[1 - .8413] = .3174.$$

CHAPTER 2

1. 2/5   2. 14/3   3. -2 4. $-\pi$

5. $(\sin 1 + \sin 2 + \sin 3)/3$.

6. $\int_{-3}^{8} x^4 d\alpha(x) = \int_{-3}^{-2+\varepsilon} x^4 d\alpha(x) + \int_{-2+\varepsilon}^{2-\varepsilon} x^4 d\alpha(x) + \int_{2-\varepsilon}^{8} x^4 d\alpha(x) = -256/5$

7. $\alpha(x) = \begin{cases} 0, x < 0 \\ 1, 0 \leqslant x < 2 \\ 0, 2 \leqslant x < 7.1 \\ -7.1, 7.1 \leqslant x < 10 \\ -5.1, 10 \leqslant x < 20 \\ 3.9, 20 \leqslant x \end{cases}$

9. $1/2$; $1/12$.   10. $5; 0$.   11. $31/8$; $226/12 - (31/8)^2$.

15. $\dfrac{df(s)}{ds} = \int_0^\infty -y e^{-sy} d_y P[X \leqslant y]$.

$E(X) = -\lim_{s \to 0+} \dfrac{df(s)}{ds}$.

$E(X^2) = \lim_{s \to 0+} \dfrac{d^2 f(s)}{ds^2}$.

16. $EZ = 1/\lambda$. $\mathrm{Var}(Z) = (1 + 2\lambda)/\lambda^2$.

CHAPTER 3

Section 3.0

1. We will find $\psi(u) = 1 - \sigma(u, \infty, \lambda + 1/A)$.

   But Corollary 4,

   $$\psi(u) = 1 - \lambda I_\alpha \left\{ \dfrac{1}{\int_0^\infty e^{-\alpha x} dP(x) - 1 + \alpha(p_1 + \lambda)} \right\}.$$

   Now

   $$\int_0^\infty e^{-\alpha z} dP(z) = \int_0^\infty e^{-\alpha z} A e^{-Az} dz = \dfrac{A}{\alpha + A}$$

and $p_1 = 1/A$. Hence

$$\psi(u) = 1 - \lambda I_\alpha \left\{ \frac{1}{A/(\alpha+A) - 1 + \alpha(1/A + \lambda)} \right\}.$$

The quantity within braces is equal to

$$\frac{\alpha + A}{\alpha[\alpha/A + \alpha\lambda + A\lambda]}.$$

By using partial fractions, this may be rewritten as

$$\frac{1}{\lambda}\left[\frac{1}{\alpha} - \frac{1}{1+\lambda A} \cdot \frac{1}{\alpha + \lambda A^2/(1+\lambda A)}\right].$$

We may now invert this to give

$$\psi(u) = 1 - 1 + \frac{1}{1+\lambda A} e^{-[\lambda A^2/(1+\lambda A)]u}, \quad u \geq 0.$$

2. One needs $A + C + E = 1$ and $B > 0$, $D > 0$, $F > 0$. One could allow $B = 0$, but that would require $A = 0$, etc.

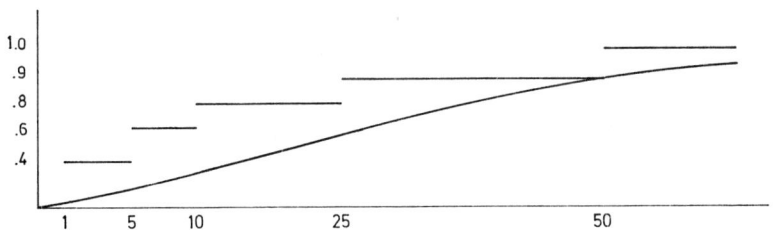

3. A Study of $P(x)$ revals that we can't hope to have $Q(1) = P(1)$. For simplicity let us solve
$Q(50) = 1 - e^{-50B} = .90$ for B. This gives $B = .046$. If we use this form for $Q(x)$,

$Q(1) = .044$, $Q(5) = .205$,
$Q(10) = .369$, $Q(25) = .683$.

A sketch of $Q(x)$ is superimposed on the above graph. For $Q(x)$, $E(X) = 1/(.046) = 21.8$, whereas for the exact distribution $P(x)$, $E(X) = 10.9$. It is merely coincidence that the one is twice the other. It is possible to improve the fit by a trial and error process of picking A, B, C, and D. For example,

$$Q(x) = 1 - .9\, e^{-.05x} - .1\, e^{-.027x}$$

preserves $Q(50) = .90$ and slightly raises $Q(5)$ to $.212$. The reader is encouraged to improve this. If he succeeds, he will also find it possible to carry through the work of exercise 4 for this improved $Q(x)$

4. $\psi(u) = \dfrac{1}{1+.046\lambda} \exp\left[\dfrac{-\lambda(.046)^2}{1+\lambda(.046)} u\right].$

5. With $\lambda = .3p_1$ and $p_1 = \dfrac{1}{.046}$, $\psi(u) = \dfrac{1}{1.3} \exp\left[\dfrac{-.3(.046)}{1.3} u\right],$

and $\psi(u) = .01$ for $u = 409.434$ units.

9. $E[X] = 1(.4) + 5(.2) + 10(.2) + 25(.1) + 50(.1)$

$\quad = 10.9.$

$E[X^2] = 1^2(.4) + 5^2(.2) + 10^2(.2) + 25^2(.1) + 50^2(.1)$

$\quad = 337.9.$

$E[X^3] = 1^3(.4) + 5^3(.2) + 10^3(.2) + 25^3(.1) + 50^3(.1)$

$\quad = 14{,}287.9.$

With $\lambda = .3p_1$,

$$\beta = \tfrac{2}{3} \dfrac{(14{,}287.9)}{(337.9)} + \dfrac{337.9}{.6(10.9)} (1-.3)$$

$\quad = 28.1896 + 36.1667 = 64.3563,$

$$\alpha = \dfrac{337.9}{.6(10.9)} \cdot (1.3)/\beta = 1.0438.$$

Now

$$\psi(u) = \dfrac{1}{1.3} \int_u^\infty \dfrac{t^{\alpha-1} e^{-t/\beta}}{\Gamma(\alpha)\beta^\alpha} dt$$

and if one uses [30], then by equation (xii) of page vii

$$\psi(u) = \dfrac{1}{1.3}\left[1 - I\left(\dfrac{u}{\beta\sqrt{\alpha}}, \alpha-1\right)\right].$$

The question is, for what u does $I\left(\dfrac{u}{65.7657}, .0438\right) = .987$? From page 2 of the above reference, $u = 65.7657(4.35) = 286.081$ units. Recall that it took 409.434 units to hold $\psi_{Q(x)}(u) = .01$.

Section 3.1

1. $E\{x(t)\} = \dfrac{d}{d\theta} \exp\{t[M(\theta)-1]\}\vert_{\theta=0}$

   $\quad = \exp\{t[M(\theta)-1]\}\, tM'(\theta)\vert_{\theta=0}$

   $\quad tp_1$ since $M(0)=1$, $M'(0)=p_1$.

   $E\{x^2(t)\} = \dfrac{d^2}{d\theta^2} \exp\{t[M(\theta)-1]\}\vert_{\theta=0}$

   $\quad = \exp\{t[M(\theta)-1]\}\, tM''(\theta)\vert_{\theta=0}$
   $\quad + \exp\{t[M(\theta)-1]\}t^2[M'(\theta)]^2\vert_{\theta=0}$

   $\quad = tp_2 + t^2 p_1^2.$

2. $P\{x(100) > 1020\}$

   $= P\left\{x(100) > p_1 t + \dfrac{20}{\sqrt{1000}}\sqrt{p_2 t}\right\}$

   $\doteq \dfrac{1}{\sqrt{2\pi}}\displaystyle\int_{-\infty}^{-20/\sqrt{1000}} e^{-y^2/2}dy$

   $\doteq \tfrac{1}{2} - \dfrac{1}{\sqrt{2\pi}}\displaystyle\int_{0}^{.632} e^{-y^2/2}dy$

   $= .264.$

   The error is $<A/10$.

3. As in #2, $x = -.632$

   $\Phi(-.632) = .2637$

   $\Phi^{(3)}(-.632) = .1962$

   $\Phi^{(4)}(-.632) = .5370$

   $\Phi^{(6)}(-.632) = 2.3054$

   $P\{x(100) > 1020\} = .2637 + \dfrac{30}{10^{\frac{3}{2}}(3!)(100)^{\frac{1}{2}}}(.1962)$

   $\quad + \dfrac{120}{100(4!)100}(.5370) + \dfrac{10(30)^2}{10^3(6!)100}(2.3054)$

   $\quad = .2637 + .0031 + .000268 + .000288 = .267356.$

   The error is $<A/1\,000$.

4. From the original equation, we see that $c=0 \Rightarrow h=0$. For our distribution, $p_1 - \int_0^\infty y e^{-y(1-h)} dy = c$, or $1 - [1/(1-h)^2] = c$. The ensuing quadratic equation in h yields $h = 1 \pm (1-c)^{-\frac{1}{2}}$, and we take the minus sign since $c=0 \Rightarrow h=0$.

5. $A_2 = \dfrac{\Gamma(\alpha)}{2\Gamma(\alpha+2)} \int_0^\infty f(z) L_2^{(\alpha)}(z) \, dz$

$= \dfrac{\Gamma(\alpha)}{2\Gamma(\alpha+2)} \{\alpha + \alpha^2 - 2(1+\alpha)\alpha + \alpha(1+\alpha)\} = 0.$

6. $E\{\exp[\theta(Y - p_1 t)]\} = e^{-\theta p_1 t} E\{\exp[\theta Y]\} = e^{-\theta p_1 t} \exp\{t[M(\theta) - 1]\}.$

Hence

$$E\{[Y - p_1 t]^i\} = e^{-t} \frac{d^i}{d\theta^i} \exp\{-\theta p_1 t + t M(\theta)\} \big|_{\theta=0}.$$

This can be done more easily by expanding the moment generating function in a series and observing the coefficients of the powers of θ.

7. Since $p_1 = 1$, $p_2 = 2$, $p_1 t = 16$, and $p_2 t = 32$, the scaling necessary to make the mean equal to the variance is one-half. After this scaling the moments of X are $\mu = 8$, $\mu_2 = 8$. Thus $\alpha = 8$, and $F(x, 16) \doteq \Gamma(x/2, 8)$. This gives the table

x	0	4	20	40
$F(x, 16)$	0	.00110	.77990	.99922

For example,

$F(20, 16) \doteq \Gamma(10, 8)$

$\doteq I(3.535, 7)$ in the terminology of [30].

Section 3.2

1. This is a slight generalization of what appears on page 417 of [28]. The nth moment

$$p_n = \int_0^\infty x^n A e^{-Ax} dx$$

$$= \int_0^\infty \left(\frac{w}{A}\right)^n e^{-w} dw = n!/A^n.$$

186

$$p^{2*}(x) = \int_0^x P(x-v)\,dP(v)$$

$$= \int_0^x [1 - e^{-A(x-v)}]\, A\, e^{-Av}\, dv$$

$$= 1 - \sum_{n=0}^{1} \frac{e^{-Ax}(Ax)^n}{n!}$$

$$= \int_0^{Ax} \frac{e^{-v}v}{1!}\, dv.$$

By induction,

$$P^{n*}(x) = \int_0^{Ax} \frac{e^{-v}v^{n-1}}{(n-1)!}\, dv,$$

the incomplete gamma function. Therefore, for $x \geqslant 0$,

$$F(x,t) = e^{-t} + \sum_{n=1}^{\infty} \frac{e^{-t}t^n}{n!} \int_0^{Ax} \frac{e^{-v}v^{n-1}}{(n-1)!}\, dv.$$

2. $P^{2*}(x) = \int_0^x P(x-v)\,dP(v) = \begin{cases} 0, & x < 2p_1 \\ 1, & x \geqslant 2p_1 \end{cases}.$

In general,

$$P^{n*}(x) = \begin{cases} 0, & x < np_1 \\ 1, & x \geqslant np_1 \end{cases}.$$

Assume that $kp_1 \leqslant x < kp_1 + 1$ for k a non-negative integer. Then

$$F(x,t) = e^{-t} + \sum_{n=1}^{k} \frac{e^{-t}t^n}{n!} = \sum_{n=0}^{[x/p_1]} \frac{e^{-t}t^n}{n!}$$

where the sum is zero if $k = 0$.

3. $H^*(x) = A \int_0^x e^{-Ay}\, dy$

$\qquad = 1 - e^{-Ax} \quad \text{for } x \geqslant 0.$

Therefore, by problem 1,

$$H_n^*(x) = \int_0^{Ax} \frac{e^{-v}v^{n-1}}{(n-1)!}\, dv.$$

Therefore,

$$\psi(u) = 1 - \frac{\lambda}{1/A + \lambda} - \frac{\lambda}{1/A + \lambda} \sum_{n=1}^{\infty} \left(\frac{1}{1+A\lambda}\right)^n H_n^*(u)$$

$$= 1 - \frac{\lambda A}{1 + \lambda A} - \frac{\lambda A}{(1+\lambda A)} \int_0^{Au} \sum_{n=1}^{\infty} \frac{1}{(1+A\lambda)^n} \frac{e^{-v} v^{n-1}}{(n-1)!} dv$$

$$= 1 - \frac{\lambda A}{1+\lambda A} - \frac{\lambda A}{(1+\lambda A)^2} \int_0^{Au} \exp\left\{v\left[\frac{-A\lambda}{1+A\lambda}\right]\right\} dv$$

$$= 1 - \frac{\lambda A}{1+\lambda A} + \frac{1}{1+\lambda A}\left[-1 + \exp\left\{\frac{-A^2\lambda u}{1+A\lambda}\right\}\right]$$

$$= \frac{1}{1+\lambda A} \exp\left\{-\frac{A^2 \lambda u}{1+A\lambda}\right\}.$$

This agrees with the result obtained through Laplace transforms in section 3.0, as it should.

Section 3.3

1. 1. Select a suitable claim distribution $P(x)$, a value t of operational time, and a value x.
   2. Select a random digit t_1 from a Poisson distribution with mean t.
   3. Select t_1 variates from $P(x)$, say $x_1, x_2, ..., x_{t_1}$.
   4. If $\sum_{i=1}^{t_1} x_i \leqslant x$, count a 1; otherwise, count a 0.
   5. Repeat steps 2, 3, 4 n times.
   6. If k ones are counted, the estimate of $F(x, t)$ is k/n, with an estimated standard error of

$$\left[\frac{k/n(1-k/n)}{n}\right]^{\frac{1}{2}}.$$

CHAPTER 4

Section 4.0

1. $P[X(\cdot) : X(0) = 0, 2 < X(1/4) \leqslant 3]$

$$= \frac{1}{\sqrt{2\pi(1/4)}} \int_2^3 e^{-x^2/2(1/4)} dx$$

$$\doteq 0.$$

2. Again zero.

3. $P[X(\cdot) : X(0) = 0, 0 < X(1/4) \leq 1]$

$$= \frac{1}{\sqrt{2\pi(1/4)}} \int_0^1 e^{-x^2/2(1/4)} dx$$

$$= .47725.$$

4. By the one-dimensional Simpson's Rule,

$$\int_b^{b+2k} \int_a^{a+2h} f(x,y)\,dx\,dy$$

$$= \int_b^{b+2k} \frac{h}{3}[f(a,y) + 4f(a+h,y) + f(a+2h,y)]\,dy.$$

Applying the Rule to each separate integral, the answer follows.

5. .295161.

6. $P[X(\cdot) : X(0) = 0, 0 < X(\frac{1}{4}) \leq 1, 0 < X(\frac{1}{2}) \leq 1, -1 < X(\frac{3}{4}) \leq 0]$

$$= \frac{1}{\sqrt{(2\pi)^3(1/4)^3}} \int_0^1 \int_0^1 \int_{-1}^0 \exp\{-2x_1^2 - 2(x_2-x_1)^2 - 2(x_3-x_2)^2\}$$

$$\times dx_3\,dx_2\,dx_1$$

$$= \left(\frac{2}{\pi}\right)^{3/2} \int_0^1 \int_0^1 \exp\{-2x_1^2 - 2(x_2-x_1)^2\}$$

$$\times \{\tfrac{1}{6}\exp[-2(-1-x_2)^2] + \tfrac{4}{6}\exp[-2(-\tfrac{1}{2}-x_2)^2]$$

$$+ \tfrac{1}{6}\exp[-2(-x_2)^2]\}\,dx_2\,dx_1$$

$$= \left(\frac{2}{\pi}\right)^{3/2} \int_0^1 \exp[-2x_1^2]\{\tfrac{1}{36}[\exp[-2x_1^2](e^{-2} + 4e^{-1/2} + e^0)]$$

$$+ \tfrac{4}{46}[e^{-2(x_1-1/2)^2}(e^{-9/2} + 4e^{-2} + e^{-1/2})]$$

$$+ \tfrac{1}{36}[e^{-2(1-x_1)^2}(e^{-8} + 4e^{-9/2} + e^{-2})]\}\,dx_1$$

$$\doteq .0696.$$

7. .9545.

8. The two dimensional Simpson's rule gives an approximate value in excess of 1.

Section 4.1.

4a. By interchanging order of integration (Fubini Theorem)

$$\int_{C_0[0,1]} x(\tfrac{1}{4})\, x(\tfrac{1}{2})\, x(\tfrac{3}{4})\, d_w x$$

$$= \int_{-\infty}^{\infty}\int_{-\infty}^{\infty}\int_{-\infty}^{\infty} u_1 u_2 u_3 \exp\left\{-\frac{(u_3-u_2)^2}{2(1/4)} - \frac{(u_2-u_1)^2}{2(1/4)} - \frac{u_1^2}{2(1/4)}\right\}$$

$$\times (2\pi)^{-3/2}[1/4]^{-3/2} du_3\, du_2\, du_1$$

$$= \int_{-\infty}^{\infty}\int_{-\infty}^{\infty} u_1 u_2^2 \exp\left\{-\frac{(u_2-u_1)^2}{\tfrac{1}{2}} - \frac{u_1^2}{\tfrac{1}{2}}\right\}$$

$$\times (2\pi)^{-1}[1/4]^{-1} du_2\, du_1$$

$$= \int_{-\infty}^{\infty} u_1(u_1^2+1/4)\exp[-2u_1^2]\frac{1}{\sqrt{\pi/2}}\, du_1 = 0.$$

Parts b and c are computed in a similar manner.

Section 4.3

1. $\displaystyle\int_{-\infty}^{\infty} g(x)\frac{e^{-x^2/(2\sigma^2)}}{\sqrt{2\pi\sigma^2}}\, dx = \int_{-\infty}^{\infty} g(\sigma y)\frac{e^{-y^2/2}}{\sqrt{2\pi}}\, dy.$

By bounded convergence,

$$\lim_{\sigma^2 \to 0}\int_{-\infty}^{\infty} g(\sigma y)\frac{e^{-y^2/2}}{\sqrt{2\pi}}\, dy = \int_{-\infty}^{\infty} g(0)\frac{e^{-y^2/2}}{\sqrt{2\pi}}\, dy = g(0).$$

Section 4.6

1. $C_1 = 1/(1-T)$, $C_2 = 0$, $f(t) = -1$,

and

$$\int_0^T d\left\{\frac{X^2(t)}{1-t}\right\} = \frac{X^2(T)}{1-T}.$$

2. $C_1 = (\beta_1/\beta_0)^{\frac{1}{2}} \exp\{T(\beta_1-\beta_0)/2\}$,

$C_2 = (\sigma_1^2 - \sigma_0^2)/(\sigma_0^2\sigma_1^2)$,

$f(t) = (\beta_0 - \beta_1)e^{-t(\beta_0+\beta_1)}/K.$

3. $K(u, v) \equiv 1$. See Section 4.0. $\sigma(K) = T$ because $\int_0^T \varphi(u)\, du = \lambda \varphi(t)$, $0 \leq t \leq T$, implies that $\varphi(u)$ is a constant which implies that $\lambda = T$. For $T < 1$, $1 \notin \sigma(K)$.

The function $k(u) \equiv 0$ since $m(t) \equiv 0$. (See Section 4.0)

Section 4.7

2. $P\{\max_{0 \leq t \leq 1} |X(t)| \leq 1\} \doteq \dfrac{4}{\pi} e^{-(\pi^2/16)} - \dfrac{4}{3\pi} e^{-(9\pi^2/16)} + \dfrac{4}{5\pi} e^{-(25\pi^2/16)} - \dfrac{4}{7\pi} e^{-(49\pi^2/16)}$

$= 1.2732\, e^{-.617} - .4244\, e^{-5.553} + .2546\, e^{-15.425} - .1751\, e^{-30.233}$

$\doteq 1.2732(.539566) - .4244(.003881)$

$\doteq .6871$

One term.

3. $P\{X(t) < t + .5,\, 0 \leq t \leq 2\} = N\left[\dfrac{2.5}{\sqrt{2}}\right] - e^{-1} N\left[\dfrac{1.5}{\sqrt{2}}\right]$

$= N[1.768] - .367879\, N[1.061]$

$= .9616 - .3147$

$= .6469$

4. $\dfrac{2}{\pi}$ arc sin $\sqrt{.25}$

$= \dfrac{2}{\pi}$ arc sin $(.5) = \dfrac{2}{\pi}\left(\dfrac{\pi}{6}\right) = \dfrac{1}{3}$.

$\dfrac{2}{\pi}$ arc sin $\sqrt{.50}$

$= \dfrac{2}{\pi}$ arc sin $(.707) = \dfrac{2}{\pi}\left(\dfrac{\pi}{4}\right) = \dfrac{1}{2}$.

$\dfrac{2}{\pi}$ arc sin $\sqrt{.75}$

$= \dfrac{2}{\pi}$ arc sin $(.866) = \dfrac{2}{\pi}\left(\dfrac{\pi}{3}\right) = \dfrac{2}{3}$.

CHAPTER 6

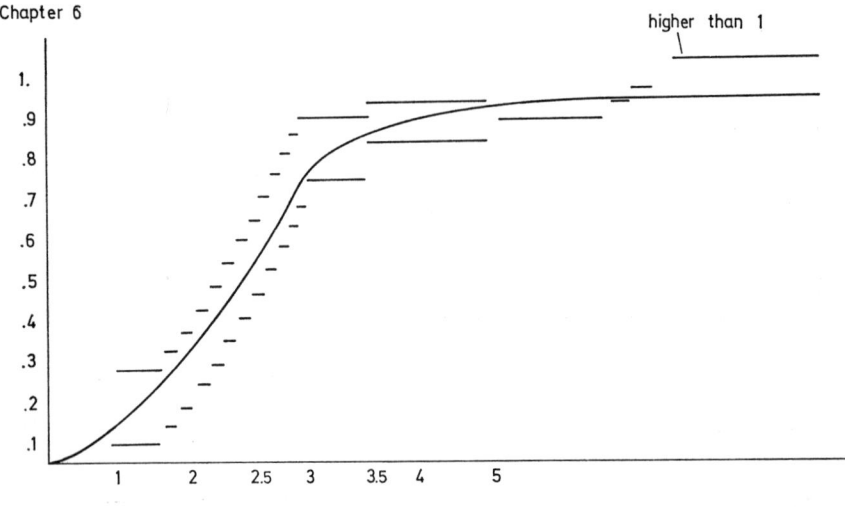

min $\{F_{35}(x)+.238, 1\}$

2. Plot $N(2.5, 1)$ and see if it rises above the upper confidence contour.

$F(2) = F^*(-.5) = .3$

$F(3) = F^*(.5) = .69$

$F(3.5) = F^*(1) = .85$, etc.

This curve has been superimposed on the above graph. It does not rise above the upper confidence contour so that we would not reject the null hypothesis.

3. For the one-sided statistics, $1.65/\sqrt{50} = .24$ for 90% and

$1.95/10 = .195$ for 95%.

For the two-sided statistic, $1.95/\sqrt{50} = .276$ for 90% and

$2.25/10 = .225$ for 95%.